MATERIALS SCIENCE AND TECHNOLOGIES

DIAMONDS

PROPERTIES, SYNTHESIS AND APPLICATIONS

Materials Science and Technologies

Additional books in this series can be found on Nova's website under the Series tab.

Additional E-books in this series can be found on Nova's website under the E-book tab.

Earth Sciences in the 21st Century

Additional books in this series can be found on Nova's website under the Series tab.

Additional E-books in this series can be found on Nova's website under the E-book tab.

MATERIALS SCIENCE AND TECHNOLOGIES

DIAMONDS

PROPERTIES, SYNTHESIS AND APPLICATIONS

TIMM EISENBERG
AND
EMILIE SCHREINER
EDITORS

Nova Science Publishers, Inc.
New York

Copyright © 2012 by Nova Science Publishers, Inc.

All rights reserved. No part of this book may be reproduced, stored in a retrieval system or transmitted in any form or by any means: electronic, electrostatic, magnetic, tape, mechanical photocopying, recording or otherwise without the written permission of the Publisher.

For permission to use material from this book please contact us:
Telephone 631-231-7269; Fax 631-231-8175
Web Site: http://www.novapublishers.com

NOTICE TO THE READER

The Publisher has taken reasonable care in the preparation of this book, but makes no expressed or implied warranty of any kind and assumes no responsibility for any errors or omissions. No liability is assumed for incidental or consequential damages in connection with or arising out of information contained in this book. The Publisher shall not be liable for any special, consequential, or exemplary damages resulting, in whole or in part, from the readers' use of, or reliance upon, this material. Any parts of this book based on government reports are so indicated and copyright is claimed for those parts to the extent applicable to compilations of such works.

Independent verification should be sought for any data, advice or recommendations contained in this book. In addition, no responsibility is assumed by the publisher for any injury and/or damage to persons or property arising from any methods, products, instructions, ideas or otherwise contained in this publication.

This publication is designed to provide accurate and authoritative information with regard to the subject matter covered herein. It is sold with the clear understanding that the Publisher is not engaged in rendering legal or any other professional services. If legal or any other expert assistance is required, the services of a competent person should be sought. FROM A DECLARATION OF PARTICIPANTS JOINTLY ADOPTED BY A COMMITTEE OF THE AMERICAN BAR ASSOCIATION AND A COMMITTEE OF PUBLISHERS.

Additional color graphics may be available in the e-book version of this book.

LIBRARY OF CONGRESS CATALOGING-IN-PUBLICATION DATA

Diamonds : properties, synthesis, and applications / editors, Timm Eisenberg and Emilie Schreiner.
  p. cm.
 Includes index.
 ISBN 978-1-61470-591-8 (hardcover)
 1. Diamonds. 2. Diamonds, Artificial. 3. Diamonds, Industrial. I. Eisenberg, Timm. II. Schreiner, Emilie.
 TN990.D54 2011
 620.1'98--dc23
                              2011024375

Published by Nova Science Publishers, Inc. † New York

CONTENTS

Preface		vii
Chapter 1	Amperometric Biosensors Based on Boron-Doped Diamond Electrodes *Yanli Zhou and Jinfang Zhi*	1
Chapter 2	Optical Properties of Diamond Like Carbon Thin Films *Sk. Faruque Ahmed*	29
Chapter 3	Polishing of diamond Surfaces *Y. Chen and L. C. Zhang*	55
Chapter 4	Electrochemical Oxidation of Organic Pollutants in Aqueous Solutions Using Boron-Doped Diamond Anodes: Cyclic Voltammetric Behavior *Nasr Bensalah, Aline S. Sales and Carlos A. Martínez-Huitle*	73
Chapter 5	Grinding Characteristics of Diamond Film Using Composite Electro-Plating in-Process Sharpening Technique *Yuang-Cherng Chiou, Rong-Tsong Lee and Tai-Jia Chen*	93
Chapter 6	Diamond Deposited by a Direct Current Hot Cathode Plasma Discharge: Characterization And Applications *M. Reinoso, F. Álvarez, E. B. Halac, and H. Huck*	113
Chapter 7	Boron Doped CVD Diamond Films Grown on Ceramic Substrates *Lívia Elisabeth Vasconcellos de Siqueira Brandão, Shay Reboh, Fabrício Casarin, Rafael Fernando Pires, Paulo Pureur Neto and Naira Maria Balzaretti*	131
Chapter 8	Specific Defects Induced by Molecular Beam Implantation into a Diamond *S. T. Nakagawa*	145
Chapter 9	Diamond and Related Materials for Biological Applications *Andrew Hopper and Frederik Claeyssens*	155

Chapter 10	Plasma Enhanced CVD Diamond Coatings on Transition Metal Substrates: An Interfacial Chemistry Study by Synchrotron Radiation *Yuanshi Li and Akira Hirose*	**195**
Index		**217**

PREFACE

This book gathers research from across the globe in the study of the properties, synthesis and applications of diamonds. Topics discussed in this compilation include amperometric biosensors based on boron-doped diamond electrodes; optical properties of diamond like carbon thin films; dynamic friction polishing of diamond surfaces; boron-doped diamond as a versatile electrode material; diamond deposition process and its application to substrates and specific defects induced by molecular beam implantation into a diamond.

Chapter 1 - Boron-doped diamond (BDD) thin films, as one kind of electrode materials, are gaining big interests. BDD electrodes outperform conventional electrodes in terms of wide electrochemical potential window, low and stable capacitive background current, high response reproducibility and long-term response stability, and good biocompatibility. Combining the superior properties of BDD electrodes with the merits of biosensors, such as specificity, sensitivity, and fast response, amperometric biosensors based on BDD electrodes have been reported by many researchers. Electrochemical reactions perform at the interface between electrolyte solutions and the electrodes surfaces, so the surface structures and properties of the BDD electrodes are important for electrochemical detection. In this chapter, the recent advances of BDD electrodes with different surfaces including hydrogen-terminated, oxygen-terminated, metal nanoparticles-modified, amine-terminated, and carboxyl-terminated thin films, and different nanostructures, for the construction of various biosensors or the direct detection of biomolecules were described. The future trends of BDD electrodes in biosensing were also discussed.

Chapter 2 - Carbon-based materials, clusters, and molecules are unique in many ways. Carbon based materials are ideally suitable as molecular level building blocks for nanoscale systems design, fabrication and applications. From a structural or functional materials perspective, carbon is the only element that exists in a variety of shapes and forms with varying physical and chemical properties. One distinction relates to the many possible configurations of the electronic states of a carbon atom, which is known as the hybridization of atomic orbitals and relates to the bonding of a carbon atom to its nearest neighbors. Diamond like carbon (DLC) can be defined as a metastable phase of amorphous carbon (a-C) or hydrogenated amorphous carbon (a-C:H) containing a significant fraction of sp^3 hybridization. As well, it is commonly accepted that the graphitic sp^2 clusters are embedded in the amorphous sp^3 bonded matrix, where the collective behavior of sp^2 sites is responsible for the optical and electrical properties, whereas the sp^3 regions govern mechanical properties. DLC has some extreme properties like diamond, such as high mechanical hardness, chemical inertness, dielectric strength, low wear and friction, optical transparency in the visible and infrared region, high electrical resistivity, high thermal conductivity and low electron affinity. These properties are promising for a wide range of applications, as protective coatings in areas, such as optical windows, magnetic storage disks, car parts, biomedical applications and as micro-electromechanical devices and electronic applications. In this chapter, the synthesis and characterizations particularly optical properties of the different metal and nonmetal doped DLC thin films have been discussed in details.

Chapter 3 - Since the successful synthesis of diamond, especially the rapid development of chemical vapour deposition (CVD) diamond technology, diamond has been used extensively in industry and the effective polishing of diamond surfaces has become increasingly important. Over the years, various techniques using a mechanical, chemical or a thermal method, or their synergistic combinations have been developed, of which the dynamic friction polishing (DFP) method is relatively new and appears to be an attractive alternative to provide the efficiency that the conventional methods cannot achieve. In addition, the DFP is an abrasive free technique, requires simple machinery, and can be implemented in a normal ambient environment. This chapter reviews the latest development on DFP of diamond surfaces, which includes polishing equipments, estimation of interface temperature, exploration of the material removal mechanism, modelling of material removal rate, establishment of polishing map for nanometric surface finish and characterization of surface integrity. It also presents the applications of the method to single crystalline diamond, polycrystalline diamond composites, and CVD diamond films.

Chapter 4 - In recent years boron-doped diamond (BDD) has emerged as a versatile and sensitive electrode material and has been used in an increasing number of applications. This article reviews some of the factors which influence its performance as an electrode, in particular focusing on cyclic voltammetric behavior during electrochemical oxidation process. The typical electrochemical response of a BDD electrode in presence or absence of organic pollutants in solution is discussed and how this relates to complexities in its process performance. The implications for the applications of BDD and prospects for the future are discussed.

Chapter 5 - To finish the CVD diamond film surface, the composite electroplating is simultaneously introduced into the grinding process for sharpening the grinder or disc (briefly called as CEPIS method). In this grinding process, the grinder of cathode and the nickel plate of anode are connected to DC power supply, and they are immersed in plating bath containing diamond particles so that metal ions with diamond particles are deposited onto a grinder in-process to expose fresh sharp particles. Results show that the removal rate of diamond film increases with increasing current density. This removal rate under the current density of 7.5 ASD is 3.8 times higher than that under 0 ASD or the well-known traditional grinding method. Moreover, the effect of bath composition on the coating structure deposited on a grinder or disc is also investigated. The real area of contact between the grinder and the CVD diamond film increases with increasing nickel chloride concentration due to the influence of the coating structure during the grinding process using a CEPIS method so that the grinder can hold the diamond particles rigidly. Consequently, the grinding ability of the grinder can be significantly improved, where a mirror-like surface of the CVD diamond film can be achieved.

Chapter 6 - A diamond deposition process is introduced and discussed together with its applications to several substrates. Diamond films are deposited by a direct current plasma discharge, assisted by electron emission from a hot cathode consisting of one or more tungsten filaments. This method allows good quality diamond film growth at a rate of up to 10 μm h^{-1} over a relatively large area (8 cm^2). This deposition method is evaluated on different substrates (diamond, silicon, steels and alumina). Several surface pre-treatments are employed for each substrate: silicon diffusion, silicon interlayers, diamond seeding, and surface scratching using diamond and SiC powder. The results are described and analyzed in detail in each case. Deposited diamonds are characterized by Raman spectroscopy, electron microscopy (SEM), energy dispersive spectroscopy (EDS), and X-ray diffraction (XRD). Potential applications are described below.

Chapter 7 - Boron doped diamond films have been grown adhered to silicon substrates by chemical vapor deposition (CVD) using boron containing gases that present high degree of toxicity. This chapter aims to investigate and discuss the boron incorporation into self-standing CVD diamond films grown on ceramic substrates. It was shown that it is possible to grow free-standing boron doped CVD diamond films on partially stabilized zirconia substrates using boron powder as the source for doping. Both surfaces of the diamond films, the one in contact with the substrate (smooth surface) and the one in contact with the plasma during deposition (rough surface), were analyzed by X-ray diffraction (XRD), Raman spectroscopy, scanning electron microscopy (SEM), transmission electron microscopy (TEM) and focused ion beam microscopy (FIB). The electric behavior of the doped films was evaluated by resistivity techniques by the 4-point probe method. For comparison, non-doped diamond films were also investigated by the same techniques. Results from Raman spectroscopy showed the Fano component, corroborating the boron incorporation into the film structure with concentration up to ~10^{20} cm^{-3}, whereas no significant change in the lattice parameter of diamond was detected by X ray diffraction. The measurement of the resistivity as a function of temperature confirmed the semiconductor behavior, as expected for p-type diamond. For both doped and undoped films, the micrographs obtained by electron microscopy showed the presence of crystalline defects like twinning and preferential orientation. However, high resolution transmission electron microscopy revealed the presence of nanometric sized clusters, approximately 2 nm size, only for the boron-doped films, probably related to the segregation of boron powder used during the diamond film deposition process.

Chapter 8 - Carbon is a leading element for innovative semiconductor devices. Here we consider one example of this; a specific defect the NV-N (nitrogen-vacancy-nitrogen) center in diamond. This is useful for the NOT logic operation for emerging quantum computers. An essential condition for these NV-N centers is that both N atoms are on a substitutional site and the intrapair distance (R_{N-N}) between them is properly separated at about 2 nm. Using a N$_2$ molecule as a projectile, we have examined what happens in a pure diamond after implantation by making use of a classical molecular dynamics (MD) and a crystallographic analysis tool called Pixel Mapping (PM). We found that sub-keV per atom energies are sufficient to dissociate the N$_2$ to an appropriate separation distance, if hot-implantation at a temperature of substrate of 900 ~ 1000 K is used. A phase transition from a crystalline to amorphous (CA) state occurred in a few ps after ion implantation. Then the damaged crystallinity starts to recover in one ns, presumably by global phonons.

Chapter 9 - Diamond and related materials, such as diamond like carbon (DLC) are currently widely studied as materials for biology, both in thin film form (for substrates of biosensors or to enhance cell growth) and in nanoparticle form (as fluorescent markers and drug delivery vehicles). This review highlights and summarises important advances in the emerging field of diamond based biomaterials. Its beneficial electrical and chemical properties such as a high refractive index, low surface roughness, biocompatibility and corrosion-resistance make diamond and its related materials suitable for a wide range of biological applications. Additionally, reliable chemical functionalisation routes of diamond surfaces have recently emerged, enabling the controllable covalent attachment of biomolecules onto this surface. Nanodiamond (ND) and ultrananocrystalline diamond (UNCD) films can be deposited under relatively mild conditions to biological surfaces as a substrate for further chemical functionalisation. Such chemical additions could pave the way for a variety of future applications such as in biosensing and microelectromechanical (MEMs) devices. Diamond films have also been suggested as a suitable method of improving the biocompatibility of objects which may come into contact with physiological fluids *in vivo* such as surgical tools and artificial prostheses. At the smaller length scale diamond nanoparticles have also been synthesised and have been suggested as suitable carriers for drug and gene delivery, achieving promising results for insulin delivery into cells. Nanodiamond particles may also be utilising for bioimaging purposes, where their small size (5 nm diameter) enables them to cross the cell membrane permitting their observation within individual cells by fluorescence microscopy. Similarly, carbon nanodots (C-dots, carbon nanoparticles with diameters typically below 10 nm) are also well suited to roles in bioimaging. Taken up by cells through endocytosis, they are thought to be potential successors to quantum dots which have limited applications for *in vivo* and *in vitro* imaging due to their inherent cytotoxicity. Their widespread advantages, such as low cost synthesis, chemical stability, ability for conjugation/functionalisation and resilience to photobleaching illustrate their potential for applications in optical imaging. The remarkable properties exhibited by diamond and its associated forms continue to appeal to scientists as research in the area constantly proliferates. It is envisaged that diamond-based materials will play an increasingly important role in the future of biomaterial development.

Chapter 10 - Diamond coating on transition metal substrates including the typical Fe, Co, Ni-based alloys, and Cu, Ti alloys has been an important research topic because it can combine the advantages of diamond coatings, which are superhard, highly thermal conductive, chemically inert and wear resistant, with those of the underlying substrates including low materials cost, superior mechanical/physical performances and excellent machinability. However, it is difficult to obtain high quality and well adherent diamond coatings on these substrates because several key technical barriers. During the deposition process, complex interfacial reaction can occur between the gaseous precursors and the substrate components due to the high reactivity, solubility or diffusivity of carbon and/or hydrogen with the metal elements. The other problem is that the huge mismatch of the thermal expansion coefficients between the diamond coating and substrate materials that cause severe delamination of the coatings produced due to accumulated internal stress. We have performed comprehensive investigations on the diamond deposition on a series of transition metal and their alloys in terms of the nucleation, growth, and adhesion ability. Different deposition performances and growth mechanism are observed. In this chapter, we will introduce our new progress on the diamond nucleation and growth on the conventional transition metals and, especially, the interfacial chemistry and fine structures of the diamond coatings and the transition metal interfaces that have been addressed by TEM and synchrotron radiation.

Chapter 1

AMPEROMETRIC BIOSENSORS BASED ON BORON-DOPED DIAMOND ELECTRODES

Yanli Zhou[1] and Jinfang Zhi[2],*

[1]Department of Chemistry, Shangqiu Normal University,
No. 298, Wenhua Road, Shangqiu 476000, China
[2]Key Laboratory of Photochemical Conversion and Optoelectronic Materials,
Technical Institute of Physics and Chemistry, No. 2, Beiyitiao,
Zhong-guan-cun, Haidian District, Beijing 100190, China

ABSTRACT

Boron-doped diamond (BDD) thin films, as one kind of electrode materials, are gaining big interests. BDD electrodes outperform conventional electrodes in terms of wide electrochemical potential window, low and stable capacitive background current, high response reproducibility and long-term response stability, and good biocompatibility. Combining the superior properties of BDD electrodes with the merits of biosensors, such as specificity, sensitivity, and fast response, amperometric biosensors based on BDD electrodes have been reported by many researchers. Electrochemical reactions perform at the interface between electrolyte solutions and the electrodes surfaces, so the surface structures and properties of the BDD electrodes are important for electrochemical detection. In this chapter, the recent advances of BDD electrodes with different surfaces including hydrogen-terminated, oxygen-terminated, metal nanoparticles-modified, amine-terminated, and carboxyl-terminated thin films, and different nanostructures, for the construction of various biosensors or the direct detection of biomolecules were described. The future trends of BDD electrodes in biosensing were also discussed.

---

* E-mail: zhi-mail@mail.ipc.ac.cn.

1. INTRODUCTION

Biosensors have been increasingly developed for many applications in environmental monitoring, food analysis, detection of biological metabolites, and clinical chemistry etc., due to their specificity, sensitivity, accuracy, and portability (Febbraio et al., 2011; Wang, 2008; Wilson and Hu, 2000). Thereinto, combing the advantages of the electrochemical techniques, amperometric biosensors have emerged as the most promising alternative for the monitoring of biologically-related species in the past few years. The development and the performance of amperometric biosensors mainly depend on the physico-chemical characteristics of the materials employed for the construction of the transducer and the methods used for the immobilization of biomolecules. Carbon-based materials including, carbon paste, porous carbon, glassy carbon, carbon nanotubes, and carbon nanofibers have been widely employed as electrochemical transducers in the field of biosensors due to simple preparation methods, large positive potential ranges, and suitability for chemical modification (Forrow and Bayliff, 2005; Maalouf et al., 2006; Kowalczyk et al., 2011; Wang and Lin, 2008). Despite their advantages, traditional carbon electrodes still suffer drawbacks, for example, electrode fouling. This problem limits their long-term stability and leads to frequent polishing or disposal of the electrode after a few uses.

Boron-doped diamond (BDD) thin films are novel carbon materials and are gaining big interests. Nowadays, the development of diamond growth by chemical vapor deposition has enabled the preparation of the BDD electrodes with different surface structures on various substrates. According to the crystal size, the BDD electrodes could be divided into polycrystalline (grain size in μm range) films and nanocrystalline films (grain size below 100 nm). Several groups, such as, Fujishima, Swain, Pleskov, Tenne, Augus Miller, Marken and Compton groups, have been studied the electrochemical properties and the electroanalytical applications of the BDD electrodes. The results have demonstrated that the BDD electrodes, possess many outstanding properties as described below (Bouambrane et al., 1996; Compton et al., 2003; Granger et al., 2000; Ivandini et al., 2004; Fischer et al., 2004): (i) Wide electrochemical potential window in aqueous electrolyte solutions: the characteristic property, which allows the BDD electrodes to be used to detect molecules oxidization at high potentials, is ascribed to the extremely low catalytic activity of BDD electrodes for both hydrogen and oxygen generation; (ii) Low and stable capacitive background current: the low electrostatic capacity minimizes the time to stabilize the background current during the amperometry, indicating the BDD electrodes superior to other electrode materials, with enhancements in the sensitivity for the low concentration detection; (iii) High response reproducibility and long-term response stability: the properties of the insensitivity to the presence of oxygen dissolved in aqueous solution and the high resistance to deactivation by fouling due to low and weak adsorption of polar molecules, make BDD film a stable electrode for electroanalysis; (iv) morphological and micro-structural stability at extreme anodic and cathodic potentials and current densities because of the high stability of sp^3-bonded carbon; and (v) good biocompatibility due to the carbon materials, etc. Therefore, these prominent properties make BDD thin film an ideal electrode substrate for amperometric biosensors.

So far, BDD electrodes have been widely employed for the construction of various amperometric biosensors or the direct detection of biomolecules. As we know, electrochemical reactions proceed at the interface between electrolyte solutions and electrodes surfaces, so the surface structures and properties of BDD electrodes are important for electrochemical detection. The present paper summarizes the biosensors based on BDD electrodes materials with different surfaces, including hydrogen-terminated, oxygen-terminated, metal (oxide)-modified, amine-terminated, and carboxyl-terminated thin films, and different nanostructures.

2. ELECTROANALYSIS AT HYDROGEN-TERMINATED BDD ELECTRODES

The surface of an as-deposited polycrystalline BDD electrode prepared by chemical vapor deposition using hydrogen gas as carrier gas is recognized to be hydrogen-terminated and its scanning electron microscopy images is shown in Figure 1a (Compton et al., 2003). A clean and homogeneous hydrogen-terminated BDD surface could be obtained by treatment of oxygen-terminated BDD surfaces with hydrogen-plasma or heating at high temperatures (800-1000 °C) under hydrogen atmosphere. The hydrogen-terminated BDD electrodes have high stability and sensitivity for analysis of a number of biological species and the performances for the detection of several selected biomolecules are summaried.

2.1. Detection of Nicotinamide Adenine Dinucleotide

The electrochemical detection of nicotinamide adenine dinucleotide (NADH) is of great interest because it is a cofactor in a large number of dehydrogenase-based biosensors. A major problem with the bare glassy carbon and other electrodes is the deactivation apidly due to the irreverible adsorption of oxidation products on the surfaces. Another problem with the unmodified and modified electrodes is the high background current due to the influence of oxygen present in the solution.

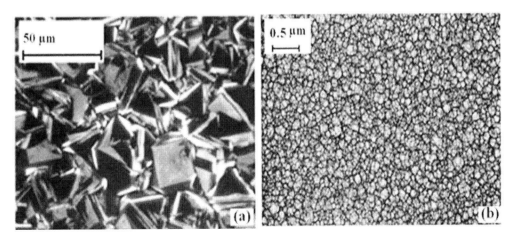

Figure 1. Scanning electron microscopy images of (a) a polycrystalline BDD film, (b) a nanocrystalline BDD film.

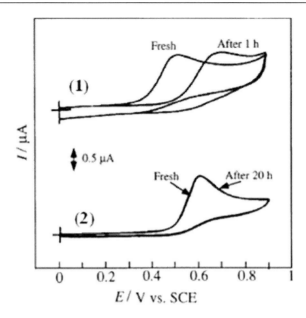

Figure 2. Cyclic voltammograms for 50 μM NADH at glassy carbon (1) and as-deposited BDD (2) electrodes in 0.1 M phosphate buffer solution (PBS, pH 7.0). The scan rate was 20 mV s^{-1}.

Fujishima and his coworkers investigated electrochemical oxidation of NADH at as-deposited BDD electrodes (Fujishima et al., 1999; Rao et al., 1999). The as-deposited BDD electrodes exhibited highly reproducible and stable response of cyclic voltammograms for its oxidation, unlike glassy carbon electrodes, at which a significant shift of ~ 200 mV in the peak potential was observed within one hour, as shown in Figure 2. A high sensitivity with a detection limit of 10 nM was also obtained. An as-deposited BDD electrode incorporating an NADH mediated dehydrogenese-based ethanol biosensor revealed good performances, indicating the feasibility of use of BDD electrodes in NADH-based biosensors.

2.2. Detection of Biogenic Amines

The analysis of biogenic amines, which are a group of naturally occuring amines derived by enzymatic decarboxylation of the natural amino acids, is essential for the early diagnosis of neurotansmission defects. Sarada et al. (2000) investigated the electrochemistry determination of histamine and serotonin in 0.1 PBS (pH 7.2) at as-deposited BDD electrodes. As shown in Figure3, well-defined cyclic voltammograms for histamine were observed at the BDD electrodes. In contrast, the voltammetric signal-to-noise (S/N) ratios obtained at glassy carbon electrodes were 1 order of magnitude lower than those obtained from BDD electrodes. The comparison of the voltammograms for BDD and glassy carbon electrodes indicated the superior behavior of diamond electrode in terms of surface inertness to adsorption and response sensitivity. By use of the flow injection analysis technique, the BDD electrodes gave a good response with a linear response range of 0.5-100 μM and a detection limit of 0.01-50 μM for histamine and serotonin at BDD electrodes, respectively.

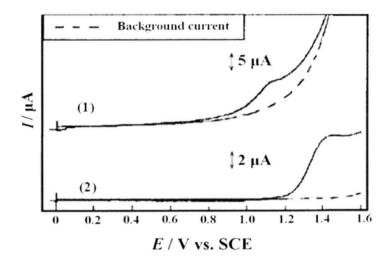

Figure 3. Linear sweep voltammograms for 100 μM histamine in 0.1 PBS (pH 7) at glassy carbon (1) and BDD (2) electrodes. The scan rate was 100 mV s^{-1}.

Another example is the electrochemical oxidation five polyamines (ethylenediamine, putrescine, cadaverine, spermine, and spermidine) at as-deposited BDD electrodes using cyclic voltammetry, hydrodynamic voltammetry, and flow injection analysis reported by Koppang et al. (1999). Well-resolved oxidation peaks with a low background signal were obtained. An oxidation model was proposed whereby the nondiamond carbon impurity sites, presumably located in the grain boundaries, generate reactive OH radicals to attack the polyamine mocules. The flow injection analysis results indicated that all five polyamines could be effectively detected at the BDD electrodes with a linear range of 1 μM – 1 mM and a detection limit of 1 μM (S/N ≈ 3-6).

2.3. Detection of L-Cysteine

L-Cysteine (CySH), a sulfur-containing amino acid, plays a crucial roles in biological systems, and its deficiency is associated with a number of clinical situations, for example, liver damage, skin lesions, slowed growth, and AIDS. Therefore, it is very important to investigate the electrochemical behavior and sensitive detection of CySH. Noble metal and bare carbon electrodes have been used for the study of CySH oxidation reaction, but in this case, the detection selectivity and sensitivity was low due to the high potential. Although high electrocatalytic activity was observed at chemically modified electrodes, the activity inevitably decreased with time.

Spătaru et al. (2001) reported the studies of the electrochemical oxidation of CySH at as-deposited BDD electrodes. Voltammetric and polarization studies showed that the CySH oxidation mechanism at BDD electrodes involved the dissociation of the proton from the thiol group, followed by the electrochemical oxidation of the CyS$^-$ species, while, at glassy carbon electrodes, the electrochemical oxidation reaction was controlled by the desorption of the reaction products. For this reason, CySH oxidation had higher stability and sensitivity at BDD electrodes compared to glassy carbon electrodes. Micromolar concentration range of 0.1-100 μM and low detection limit of 21 nM (S/N = 3) were obtained for CySH detection at the BDD electrodes. The deactivation of the electrode with time, which was common at the chemically modified electrodes, was also avoided due to the extreme electrochemical stability of the BDD electrode. That is to say, the use of BDD electrodes results in a simple and useful analytical procedure for the detection of CySH.

2.4. Detection of Oxalic Acid

Sensitive detection of oxalic acid in biological materials, i.e., urine, is required, because the level oxalic acid is an important indicator for the diagnosis of kidney stone formation. Direct electrochemical detection of oxalic acid is difficult due to a very large overpotential for the oxidation (C-C bond clevage) at conventional electrodes. Up to now, there is no report of direct oxalic acid oxidation at bare carbon electrodes, although some metal nanoparticles-modified glassy carbon electrodes has been reported. Ivandini et al. (2006) reported that oxalic acid could be electrochemicaly detected at as-deposited BDD electrodes. The BDD electrodes exhibited well-defined peaks of oxalic acid oxidation, as shown in Figure 4, with a linear response range of 0.05-10 μM and a detection limit of ~ 0.5 nM (S/N = 3). For comparison, at a glassy carbon electrode, an ill-defined peak and a high background signal were observed.

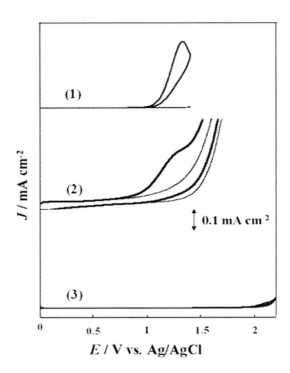

Figure 4, Cyclic voltammograms of 0.1 M PBS (pH 2.1) at the as-deposited (hydrogen-terminated) BDD (1), glassy carbon (2), and oxygen-terminated BDD (3) electrodes with (bold line) and without (thin line) the presence of 100 μM oxalic acid. The scan rate was 100 mV s^{-1}.

Clearly, the as-deposited BDD electrode showed higher sensitivity than the bare glassy carbon electrode for the detection of oxalic acid because of its low and stable background current. Moreover, at an oxygen-terminated BDD electrode, no peak was observed within the cycling potential from 0 to 2.2 V (vs. Ag/AgCl), which could be ascribed to the repulsion between the negative electrode surface and the negatively charged molecules due to its two carboxyl functional groups, suggesting that surface termination contributed highly to the control of the electrochemical reaction.

2.5. Detecction of Glucose

Sensitive, selective, reliable and fast monitoring of glucose has become a major concern troughout the world, especialy for the control and treatment of diabetes. There are two major types, i.e., enzymatic and nonenzymatic sensors, for the electrochemical detection of glucose. Although the glucose oxidation utilizing glucose oxidase (GOx) has high selectivity, the glucose biosensors based on GOx suffer from difficulties in miniaturization and instability. Therefore, the electrochemical determination of glucose without employing an enzyme has attracted increasing interest due to its stability, simplicity, reproducibility and being free from oxygen limitation. Metal electrodes including gold, platinum, copper and nickel are known for showing the electrocatalytic oxidation of glucose. However, in real biological matrices, a major problem for direct electrochemical detection of glucose is the coexistence of many interfering compounds, for example, uric acid (UA) and ascorbic acid (AA). These compounds can be oxidized at a close potential to that of glucose, which results in the overlap of voltammetric response at above metal electrodes.

Lee and Park (2005) reported that bare BDD electrodes annealed with a hydrogen flame can be used to detect glucose directly without any modification with enzymes or metallic cayalysts. Figure 5a showed a series of cyclic voltammograms for the oxidation of glucose at the hydrogen-terminated BDD electrodes and the oxidation peak current (0.65 V vs. Ag/AgCl) kept on increasing with the increasing concentration of glucose. An anodic peak was also observed during the reverse scan, which indicated that glucose was strongly adsorbed on the electrode surface, and was continuously oxidized during the reverse scan. The formation of fouling from the oxidized glucose on the electrode prevented further oxidation of glucose. However, this passive film could be easily removed by simply rinsing the electrode with deionized water. Thus, the hydrogen-terminated BDD electrode gave a linear reponse range of 0.5-10 mM for glucose, which well encompasses the physiological range of 3-8 mM. Importantly, the selective detection of glucose in the presence of AA (Figure 5b) or UA (not shown here), could be obtained at the hydrogen-terminated BDD electrode.

On the other hand, compared with the polycrystalline BDD electrodes, nanocrystalline BDD electrodes have a smoother surface while maintaining the intact BDD properties. Zhao et al. (2009a,b) reported the direct detection of glucose at the nanocrystalline BDD electrodes. The glucoe concentration is linear in the range 0.25-12 mM, indicating the possibility for diabetes diagnosis. Therefore, the superior performance of the BDD electrode may bring it as a future glucose detector in the medical field.

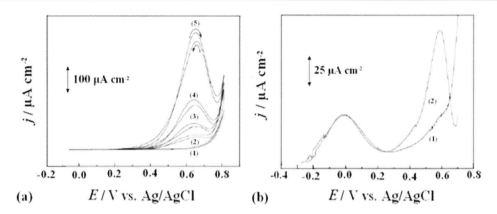

Figure 5. (a) Cyclic voltammograms of (1) 0, (2) 0.5, (3) 1.0, (4) 2.0, and (5) 5.0 mM glucose in 1.0 M NaOH at hydrogen-terminated BDD electrodes. The scan rate was 20 mV s^{-1}; (b) Square wave voltammograms of (1) 1.0 mM AA and (2) also containing 5.0 mM glucose in 1.0 M NaOH at hydrogen-terminated BDD electrodes. The pulse height was 25 mV, and the frequency was 10 Hz cm^{-2}.

2.6. Detection of Amino Acid

Analysis of amino acids has been attracted more and more attentions because amino acids, as the rudimental units of enzymes and proteins, are so closely related to life. Several research works on the electroanalytical determination of some amino acids, including tryptophan (Trp) and tyrosine (Tyr) have been described at the traditional electrodes. Unfortunately, it is still some difficult to realize the simultaneous of the above two kinds of amino acids by direct voltammetry because oxinative peaks of amino acids could not be completely separated. The simultaneous measurement of Trp and Tyr was exploited at the hydrogen-terminated BDD electrodes (Zhao et al., 2006). The oxidative peaks of these two kinds of amino acids could be completely separated at BDD electrodes by the use of differential pulse voltammetry. The peak separation of Trp and Tyr was developed to be 0.64 V when the buffer solution with the optimized pH 11.2 was employed. The alkaline medium was selected because the Trp (isoelectric pH = 5.89) and Tyr (isoelectric pH = 5.68) are negatively charged, and the surface hydrogen-terminated BDD is almost positively charged. In the presense of other amino acid, the BDD electrode gives a linear response in the range from 20-1000 μM for the two amino acids, and detection limits are 10 and 1 μM for Trp and Tyr, respectivly.

In addition, very little work has been reported previously on the detection of Tyr derivatives including phosphorylated Tyr (Tyr-P) and sulphated Tyr (Tyr-S), probably due to their high oxidation potentials. Chiku et al. (2010) studied the electrochemical oxidation of Tyr and above Tyr derivatives on the hydrogen-terminated BDD electrodes. Cyclic voltammetry for Tyr, Tyr-P and Tyr-S showed well-defined oxidation peaks at 0.8, 1.4, and 1.7 V, respectively. The peak currents show a linear function in the concentration range of 10-100 μM for Tyr and Tyr-P, and between 100-1000 μM for Tyr-S. Therefore, the Tyr and Tyr derivatives were successfully detected by an electrochemical method using BDD electrodes.

2.7. Detection of Haemoglobin

Haemoglobin (Hb), a heme protein that consists of a four-polypeptide chain, each with one heme group, can store and transport oxygen in the red blood cells. Study on the electrochemical behaviour of heme proteins is important for the fundamental understanding of their biological activity. Unlike some other small heme proteins such as cytochromes, it, however, is difficult for Hb to exhibit heterogeneous electron transfer in most cases, which means that the electron transfer of Hb is very slow. Nekrassova et al. (2004) reports the first ever direct oxidation of reduced form of human Hb at the bare BDD electrode (hydrogen-terminated). Under moderately alkaline conditions, a well defined single oxidation wave is observed. This also can be explained by the interaction of negatively charged Hb (pK_a about 8.0) and the positively charged electrode. The detection limit of 0.4 µM was obtained because the BDD electrode could overcome the problem of fouling of the electrode surface by the protein, which is a common difficulty at the solid electrode substrates.

2.8. Detection of DNA and Purines

Detection of underivatized DNA, which help to understand genetic information and related diseases, is highly desirable in order to avoid contamination problems and any sample handling losses. Direct electrochemical detection of purine bases in nucleic acids provides a simple, fast and economical method without any derivatization or hydrolysis steps. However, the electrochemical signals of the purine bases at traditional electrodes are not sensitive and reproducible because of the high oxidation potentials, where oxygen evolution current interferes.

Prado et al. (2002) demonstrated the possibility of detection of underivatized nucleic acids at as-deposited BDD electrodes using cyclic voltammetry and square wave voltammetry. Ivandini et al. (2003) further investigated the electrochemical performances of oxidation of underivatized-nucleic acids in terms of single and double stranded DNA at BDD electrodes. At an as-deposited (hydrogen-terminated) BDD electrode, at least two well-defined voltammetric peaks were observed for both single and double stranded DNA. BDD electrode was the first material to show the well-defined peaks due to its wide potential window. Linear range of 0.1-8 µg mL^{-1} and low detection limits for both guanine and adenine residues at as-deposited BDD electrodes were obtained. The results shows the possibility of using BDD film as an electrochemical sensor for direct detection of DNA molecules.

In addition, Wang et al. (2004) investigated the measurements of purines and related compounds on a microchip capillary electrophoresis coupled with a BDD electrode detector. As we know, the separation and detection of purine bases and purine-containing compounds represents a challenging task due to the crucial roles of these compounds in a variety of biochemical processes. A rapid and effective separation and favorable S/N characteristics at high detection potential (+1.3 V) for the measurement of five pruines (guanine, hypoxanthine, guanosine, xanthine, and uric acid) were obtained at the BDD end-channel amperometric detector. The high speed, negligible sample consumption, low power requirements and small size of the capillary electrophoresis microchip device based on using the BDD electrode as a detector offer an interesting candidate for determination of purine related compounds in clinical, biotechnological, and food matrics.

In a word, hydrogen-terminated BDD shows advantages for electrochemical oxidation of a broad range of biological compounds, especially negative charged molecules (NADH, L-cysteine, oxalic acid, glucose, Tyr, Trp, Hb, and DNA in the given buffer solutions), based on their wide potential window and low background current. In most of these studies, BDD electrodes was found to outperform glassy carbon in terms of stability and sensitivity. Fundamental studies on BDD films have revealed that the lack of oxygen functional groups on the as-deposited BDD surface and the very low tendency for adsorption of most chemical species on the inert surface of diamond are mainly responsiple for the superior performance of the BDD electrodes. Additionally, the oxidation reaction of the above bio-molecules is very less at an oxygen-terminated BDD electrode because of the electrostatic repulsion between its carbon-oxygen dipoles and the negative charged compounds. However, at the hydrogen-terminated surface the positive dipolar field created attracts the above bio-molecules, facilitating the electrochemical reaction. Therefore, the control of surface termination is important for the electrochemical detection of some negative charged molecules by the use of BDD electrodes.

3. ELECTROANALYSIS AT OXYGEN-TERMINATED BDD ELECTRODES

A hydrogen-terminated BDD electrodes can be altered to oxgen-terminated by a variety of methods, e.g. anodic treatment at highly positive potentials, oxgen plasma treatment, high treatment temperatures (300-1000 °C) under O_2 atmosphere, boiling in strong acid, and oxidation by strong oxidant. Electrochemical oxidation is convenient and popular for the use of electrochemistry, because the treantment can be performed in the same system as the electrochemical measurements. After treatment, a certain amount of C-OH and C=O groups were introduced on BDD surface. Compounds with positive charge maybe easily oxidized at oxygen-terminated BDD electrode in comparison with that at as-deposited BDD electrode since electrostatic attraction force from carbon-oxygen functionalities. While, as discussed above, negative charge compounds were more clearly obtained at as-deposited BDD electrode than those at oxygen-terminated BDD electrode due to the electrostatic repulsion at the oxidized surface existed. Thus, oxygen-terminated BDD electrodes could show selectivity (either enhancement or suppression) for detection of some biomolecules due to the carbon-oxygen dipoles of surface.

3.1. Selective Detection of Dopamine in the Presence of Ascorbic Acid

Dopamine (DA) is an important neurotransmitter in mamalian central nerous system. A loss of DA-containing neurons may result in serious disease such as Parkinson's disease. The electrochemical detection of DA at carbon-based electrodes has received intense attention in part due to the possible future development of practical in vivo sensors. Major problem of electrochemical detection DA is the overlapping of oxidtion waves between DA and interfering compounds (e.g., AA) at the above bare electrodes.

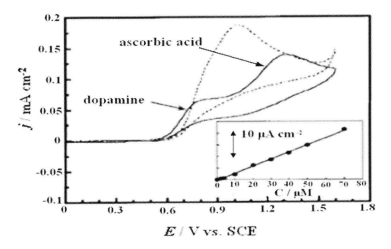

Figure 6. Cyclic voltammogram for a 0.1 M HClO₄ solution containing 0.1 mM DA and 1 mM AA at as-deposited and oxidized BDD electrodes. The scan rate was 100 mV s⁻¹. Inset shows a cyclic voltammogram calibration curve for DA in 0.1 M HClO₄ solution in the presence of 1 mM AA.

Recently, some nano-materials, e.g. carbon nanaotubes (Zhang et al., 2005), gold nanoparticles (Zhang and Jiang, 2005), and polymer thin-film (Lin et al., 2007) modified glassy carbon electrodes were used to carry out the simultaneous detection of DA and AA. However, there are difficulties remaining to be solved, such as complicatied preparation methods and long-term stability.

Fujishima's group researched the electrochemical detection of DA in the presence of AA at BDD electrodes (Popa et al., 1999; Tryk et al., 2007). As shown in Figure 6, at as-deposited BDD (hydrogen-terminated) electrode, the anodic peak potential ($E_{p,a}$) of DA and AA was 0.76 and 0.80 V (vs. saturated calomel electrode, SCE) in 0.1 M HClO₄, respectively. After electrochemical treatment of BDD (oxygen-terminated) electrode, the $E_{p,a}$ of DA and AA was 0.80 and 1.3 V (vs. SCE), respectively. Cyclic voltammogram for a solution containing both DA and AA in 0.1 M HClO₄ at the oxygen-terminated BDD electrode exhibited two well-defined anodic peaks. Therefore, DA can be selectively determined in the presence of AA using the oxygen-terminated BDD electrode, and as a result, a low detection limit of 50 nM (S/N = 3) was obtained.

Possible explanation of the selective detection for DA is that the oxidized BDD surface acquires surface dipoles as a result of introducing C=O functional groups, which then electrostatically repel the oxygen-containing group on AA with strong dipoles. That is to say, the AA oxidation is impeded owing to the high potential required and the spatial locations separates from oxidiazed BDD surface. For the protonated DA, the interaction between the ammonium group of DA is relativly strong with both hydrogen- and oxygen-terminated BDD electrode, so that the equilibrium distances and the electron transfer rates are not greatly different.

3.2. Selective Detection of UA in the Presence of AA

UA is a very improtant biomolecule owing to its abnormal levels, which are symptoms of several diseases, such as, gout, hyperuricemia, and Lesch-Nyhan syndronme. The major obstacle in monitoring this compound is the interference from AA coexistence, which undergoes oxidation at the most same potential as UA at some conventional electrodes. Various functionalized electrodes have been attempted to separate the overlapping anodic peak of UA and AA. Although these voltammetric techniques are more selective, less costly and less time-consuming than other techniques, the reproducibility and stability is low due to the adsorption phenomena.

Popa et al. (2000) reported the selective detection of UA in the presence of AA at anodized BDD electrodes. The C=O groups of BDD electrodes was expected to form a surface dipolar field, which repeled AA molecules that have oxygen-containing functional groups surrounding a central core, as noted in the section of 3.1. In differential pulse voltammograms, the oxidation peak potential of AA was ~ 450 mV more positive than that of UA at oxized BDD electrodes. Thus, the detetion of UA could obtain excellent selectivity and sensitivity with a detection limit of 1.5×10^{-8} M (S/N = 3) in the presence of high concentrations of AA by use of chronoamperometry. Meanwhile, the oxized BDD electrode exhibted high reproducibility and long-term stability (repeatedly used for more than 3 months), and the practical analytical utility was demonstrated in human urine and serum.

3.3. Simultaneous Detection of Oxidized and Reduced Glutathione

Glutathione (GSH) is among the most important antioxidant in celles involved in enzymatical reduciton of hydroperoxides and nonenzymatical retainment of vitamin E and C in reduced and functional forms. In the presence of oxidants, GSH is oxidaized to glutathione disulfide (GSSG), which is either reduced enzymatically by glutathione reductase or excreted from cells into extracellular fluids. Thus, the GSSG/GSH ration serves as an indicatior of oxidative stress. The early techniques for the measurement of GSH and GSSG were based on an enzymatic recycling and high performance liquid chromatography method, which suffer from low sensitivity and complicated pretreatment. Electrochemical method is simple and sensitive, but the electrochemical detection of GSSG, which is a positively charged molecule, is difficult due to the high oxidation potentials.

Terashima et al. (2003a) studied the amperometric detection of GSSG and GSH at anodically oxidized BDD electrodes. Figure 7 showed a comparison of cyclic voltammograms for GSSG obtained at oxidized BDD, glassy carbon, and Pt electrodes in acidic medium (pH = 2). At the Pt electrode, a small oxidation current was observed in the potential as the peak of surface oxide formation, and the peak current overlapped with oxygen evolution when increasing potential. While a low S/N current was obtained for GSSG oxidation at glassy carbon electrodes due to surface fouling of adsorbed reaction products and the oxidation of glassy carbon surface at high potential. For comparison, the response of GSSG at oxidized BDD electrodes showed two well-defined peaks with diffusion controlled limiting currents. However, the GSSG oxidation was not so sensitive at as-deposited BDD electrodes. These was because, at the oxidized BDD electrode, the GSSG oxidation was ascribed to the electrostatical attraction between the positively charged GSSG molecule in acidic solution and a negative dipolar field formed by oxygen functional groups of electrode surface.

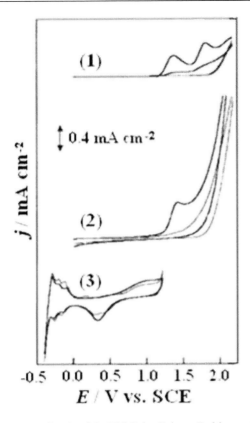

Figure 7. Cyclic voltammograms for 1 mM GSSG in Britton-Robinson buffer solution (pH 2) at oxidized BDD, glassy carbon, and Pt electrodes. The scan rate was 100 mV s^{-1}. Thin lines represent background current.

To obtain the simultaneous quantification of GSSG and GSH, liquid chromatography coupled with the oxidized BDD electrodes as electrochemical detection was used. The detection limits for GSH and GSSG were 1.4 and 1.9 nmol (S/N = 3), respectively. The variation in day-to-day reponse during 3 days were less than 3 %. The high stable amperometric response of the oxidized BDD electrode could be ascribed to the high electrochemical stability and the suface reactivation of oxidized electrodes as a detector.

3.4. Simultaneous Detection of Purine and Pyrimidine

Ivandini et al. (2007) described the simultaneous detection of purine and pyrimidine bases at BDD electrodes. At anodized BDD electrode, well-defined anodic peaks were observed for the oxidation of purine and pyrimidine bases in acid medium due to its high overpotential of oxygen evolution reaction, whereas the oxidation response of thymine was overlapped with the anodic current of oxygen evolution at as-deposited BDD electrodes. By the use of high performance liquid chromatography, the chromatograms of adenine, guanine, cytosine, thymine and 5-methlcytosine mixture were well resolved, and the electrochemical detection at anodically oxidized BDD electrode showed an order of magnitude higher sensitivity and stability than those at conventional electrodes.

In brief, the oxygen-terminated BDD electrodes have outstanding features in a much wider potentails window, and higher surface stability from fouling compared to hydrogen-terminated BDD electrodes. Importantly, the oxygen-terminated BDD electrodes are able to achieve selective detection of certain compounds under certain conditions. The significant advantages including very ease of preparation, very high stability, high sensitivity and good selectivity, make oxygen-terminated BDD electrodes an intersting candidate for the study of the direct detection of biological molecules.

4. ELECTROANALYSIS AT METAL (OXIDE)-MODIFIED BDD ELECTRODES

Assembly of ordered metal (oxide) nanoparticles, in particular gold nanoparticles, have a wide range of applications in electronics, catalysis, and analysis. Metal (oxide) nanoparticles could be immobilized on diamond surface by vacuum vapor deposition, electrochemical deposition, sputtering or layer-by-layer self-assembly. Considering the outstanding properties of BDD electrodes, assembly of ordered metal (oxide) are expected to have a wide range of applications in the field of biocatalysis.

4.1. Detection of Hydrogen Peroxide

Detection of hydrogen peroxide (H_2O_2) takes a great part in the reported amperometric biosensors as H_2O_2 is an important analyte in many fields, including industry, clinical medicine and environment. Electrochemical sensing of H_2O_2 has been done based on both of its direct oxidation at electrode surface and reduction by peroxidase. However, the direct electrochemistry of H_2O_2 requires a high over-potential. Chemical modification of the electrode surface is a well-established strategy for achieving wide applicability of the electroanalytical methodology. As we know, some metal (oxide) have electrocatalytic properties for H_2O_2 oxidation. Terashima et al. (2003b) demonstrated the electrocatalytic activity of hydrous iridium oxide (IrO_x) modified on BDD electrodes by electrodeposition. Highly dispersed and stable IrO_x nanoparticles could be obtained by the control of the deposition conditions, and the deposition of a low amount of IrO_x (ca. 2 nmol cm^{-2}) on the BDD electrode exhibited an excellent analytical performance for H_2O_2 detection with a wide dynamic range (0.1-100 μM). The sensitivity for H_2O_2 was ca. 10 times higher than that of the usual Pt-bulk electrodes. Furthermore, the IrO_x modified BDD electrodes enabled H_2O_2 detection in neutral media, which was an advantage compared to sensors obtained by IrO_x deposition on other substrates. Ivandini et al. (2005) also reported the electrochemical detection of H_2O_2 at the Pt-modified BDD electrodes prepared by implantation method at 750 KeV Pt^{2+} with a dose of 5×10^{14} cm^{-2}. The Pt-modified BDD electrode exhibited high catalytic activity with a low detection limit of 30 nM (S/N = 3) and excellent electrochemical reproducibility with a relative standard deviation of 2.91% (n = 20). The above results revealed that the metal modification on BDD thin films could be a promising method for controlling the electrochemical properties of the electrodes.

4.2. Detection of DA

To determine DA with high sensitivity and selectivity, the modification of metal nanoparticles on BDD electrode was studied. Weng et al. (2005) reported that BDD electrode was modified with gold clusters by electrodeposition and the electrochemical performance of DA and AA was investigated on the Au/BDD electrode. Gold clusters with a size distribution between 20 and 400 nm were deposited on the BDD surface after deposition. The Au/BDD electrode showed a higher activity for DA oxidation than AA; the oxidation peak of DA shifted to a less-positive potential (0.11 V) than that of AA (0.26 V), and a much higher peak current could be observed for DA oxidation than that of AA. As a result, the Au/BDD electrode could selectively determine DA in the presence of a large excess of AA with a detection limit of 0.1 µM, but it could not resist fouling. After the further modification of carboxylic groups on the Au clusters, the carboxyl- terminated Au/BDD electrode had a better antifouling effect and higher sensitivity for DA detection due to the electro-static attraction. The minimal detection limit of 1.0 nM and linear range of 10 nM to 10 µM could be achieved. The detection limit was lower than that of direct detection of DA at oxygen-terminated BDD electrodes discussed in section of 3.1, indicating that the gold clusters introduced on the surface are indeed a useful catalyst to the DA oxidation.

In addition, three-dimensional multilayer-coated PS (polystyrene) spheres were formed by a layer-by-layer self-assembly of negatively charged gold nanoparticles and a polyelectrolyte (PE) on a PS colloid (Wei et al., 2008). The Au/PE/PS multilayer spheres were modified on BDD electrodes and the modified BDD electrodes were used to study the electrochemical behaviors of DA and AA. The prepared sphere-modified BDD electrode exhibit high electrocatalytic activities toward the DA oxidation, while the AA oxidation showed almost no response and the unoxidized AA would not cause any side reactions due to electrostatic repulsion. The results displayed an effective detection of DA in the presence of AA at the modified BDD electrode with detection limit of DA of 0.8 µM.

4.3. Detection of Glucose

A nickel modified BDD electrode was fabricated by the electrodeposition from a 1 mM Ni $(NO_3)_2$ solution (pH 5), followed by repeat cycling in KOH (Toghill et al., 2010). The nickel microparticle radius of 520 nm over a range of 0.05-1.9 µM and the average number of particles with 1.2×10^6 particles cm^{-2} were obtained. The Ni-BDD electrode was used in the nonenzymatic determination of glucose in alkaline solutions. The detection of limit for glucose was 2.7 µM at the Ni-BDD electrode. Compared with the BDD electrodes of section 2.5, the lower detection of limit on the Ni-BDD electrode was ascribed to the electrocatalysis of the oxidation of glucose by the $Ni(OH)_2$/NiOOH redox couple.

In brief, the modification of metal (oxide) nanoparticles on the BDD electrode makes it a candidate to prepare highly active electrode for the catalytic oxidation (or reduction) of biomolecules. Especially, the immobilization of charged nanoparticles on BDD electrode could determinate special biomolecules in biological system with high selectivity and high sensitivity. The metal nanoparticle modification of BDD offers a simple yet effective approach to enhancing the electroanalytical ability of the electrode material.

5. BIOSENSORS BASED ON AMINE-TERMINATED BDD ELECTRODES

Due to the inert nature of BDD surfaces, the immobilization of biomolecules on BDD surfaces for the fabrication of amperometric biosensors requires surface activation procedures to provide the reactive groups, such as amino and carboxylic groups. Generally, amine-terminated BDD electrodes could be achieved by modification several methods, e.g. (i) etching a hydrogen-terminated BDD surface by NH_3 plasma in specific reactor, (ii) chemical modification of an oxidized BDD surface with (3-aminopropyl) triethoxysilane (Notsu et al., 2002), (iii) photochemical reaction of amino molecules containing a vinyl group with a hydrogen-terminated BDD surface by free radical mechanism (Yang et al., 2002) , and (iv) diazonium functionalization of 4-nitrobenzendiazonium tetrafluoroborate with hydrogen-terminated BDD electrodes by combined chemical and electrochemical processes (Yang et al., 2005). Thus, a layer of amine groups introduced on the BDD surface could serve as binding sites for attachement of biomolecules.

On the other hand, tyrosinase, also known as polyphenol oxidase, can catalyse two reactions: ortho-hydroxylation of phenols to catechol and the further oxidation of catechols to *ortho*-quinones, both in the presence of molecular oxygen. Then the quinone generated could be reduced by the electrodes, reforming the original phenols, thus forming a bio-electrocatalytic amplification cycle. Moreover, the polymerization of phenols and interferance from oxidizable species could be prevented because of the low potential for reduction of *o*-quinones on tyrosinase-based sensors. The key aspects in the construction of this kind biosensors is the choice of method and substrate for immobilizing the tyrosinase on the electrode surfaces. Tyrosinase-based biosensors reported have employed conventional electrode materials as substrates, such as glassy carbon, carbon-paste, gold, and other materials. But these enzyme electrodes suffer from complication in manipulation, desorption of enzyme from electrode materials, and/or weak in retaining the bioactivity of tyrosinase. Thus, biosensors based on negatively charged tyrosinase (pI 4.5) immobilized on the positively charged amino groups of BDD electrodes in neutral buffer solution are of considerable interest due to the outstanding properties of BDD films and the reliability of covalent immobilization.

Notsu et al. (2002) reported the immobilization of tyrosinase on a BDD electrode. Firstly, (3-aminopropyl) triethoxysilane was used to modify BDD electrode treated by electrochemical oxidation, ant then a tyrosinase film cross-linked with glutaraldehyde. The low limit with 10^{-6} M for bisphenol-A was achieved at the enzyme electrodes by using a flow injection system. However, the tyrosinase-modified BDD electrode retained its initial activity only for a few days in storage under dry conditions, due to weak bonding of (3-aminopropyl) triethoxysilane with BDD surface.

To improve the stability of tyrosinase-based BDD electrodes, Zhi's group studied the covalent immobilization of tyrosinase onto the amine-terminated BDD electrode (Zhou and Zhi, 2006; Zhou et al., 2007) . The amine active BDD surfaces were obained by two methods. One was that the hydrogen-terminated BDD surface was treated with allylamine by photochemical reaction. Another was that the hydrogen-terminated BDD surface was treated with 4-nitrobenzenediazonium tetrafluoroborate.

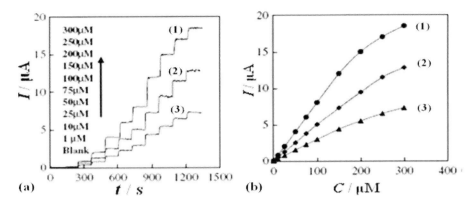

Figure 8. (a) Current-time recording and (b) calibration curve of the tyrosinase-modified BDD electrode to *p*-cresol (1), 4-chlorophenol (2), phenol (3). Supporting electrolyte was 0.1 M PBS (pH 6.5). Applied potential was -0.15 V vs. SCE.

The difference of these two methods was that the photochemical reaction linked the amino group to BDD surface via an alkyl chain, while the diazonium method typically used an aromatic ring. Both the two tyrosinase-modified BDD electrodes by the above methods exhibited fast response, high sensitivity and wide linear range for the detection of phenolic compounds, as shown in Figure 8 (taking diazonium method as example). For comparison, the sensitivity of the enzyme electrode by diazonium method was higher than the enzyme electrode by photochemical method, which might be because, the aryl ring of diazonium molecule was more conductive than alkyl chain. That is to say, the aryl ring could be well-suited for the electron transfer between enzyme molecules and the BDD electrodes. The sensitivity of the enzyme electrode by diazonium method was also higher than those of the tyrosinase biosensors reported (Sanz et al., 2005; Wang et al., 2000; Rajesh et al., 2004), which could be attributed to the high and reliable loading of enzyme by present method. The two developed enzyme electrodes could retain about 90% of its initial activity for the response of phenols after 1 month. The high stability could be ascribed to the strongly covalent bonding of tyrosinase to the BDD electrodes and the high chemical and eletrochemical stability of the BDD substrates. To sum up, the amine-functionalized BDD electrodes is an intersting alternative for application in biosening technology.

6. BIOSENSORS BASED ON CAYBOXYL-TERMINATED BDD ELECTRODES

6.1. Cytochrome *c* Biosensor

Direct electron transfer reactions between redox proteins and electrode surfaces has attacted condiderable interst. Understanding of these reactions can establish a fundamental and desirable model for studying the natural redox properties of the protein, the interfacial charge transfer process, and the relationship between their stuctrure and biological functions. Moreover, the study of direct electron transfer between proteins and underlying electrodes could also provide a platform for fabricating biosensors. It is known that the heme-containing proteins such as horseradish peroxidase (HRP), cytochrome *c*, and myoglobin, have a direct electrochemical behavior for its redox center of Fe$^{III/II}$ at the electrode surface and the ability to electrocatalyze the reduction of H$_2$O$_2$. However, direct electrochemistry of these proteins at a bare eletrode is difficult because of its extremely slow electron transfer kinetics at the electrode/solution interface and its short-live and transient response on a metal electrode surface. Common electrode material employed should be modified with mediators by special methods.

Figure 1b shows the scanning electron microscopy images of nanocrystalline BDD electrodes (Zhou et al., 2008). In addition to boron doping, the sp^2-bonded carbon phase on grain boundaries of nanocrystalline BDD films can provide charge carries and high carrier mobility pathways, which may lead to better reversible properties of electrodes for redox systems (Bennett et al., 2004). Thus, BDD electrode is expected to be a more suitable candidate to provide a high activity for biosensors. In fact, surface smoothness of BDD electrodes could affect their amperometric response significantly. It has been demonstrated that mechanical polishing of a polycrystalline BDD electrode to a nanometer-scale finish has shown to result in a well-defined voltammetric response of cytochrome *c* in solution (Marken et al., 2002). Haymond et al. (2002) have also reported that direct electron transfer could occur between cytochrome *c* existing in solution and the as-deposited nanocrystalline BDD electrode with quasi-reversible, diffusion-controlled electron transfer kinetics.

Zhi's group reported the functionalization of nanocrystalline BDD films via photochemical reaction with undecylenic acid methyl ester and subsequent removal of the protection ester groups to produce a carboxyl-terminated surface (Zhou et al., 2008). Then cytochrome *c* was successfully immobilized on nanocrystalline BDD electrode by the bonding of negatively charged carboxylic groups and positively charged lysine residue of Cyt *c* (pI 10). The cytochrome *c*-modified nanocrystalline BDD electrode showed a pair of quasi-reversible redox peaks with a formal potential (E^0) of 0.061 V (vs Ag/AgCl) in 0.1 M PBS (pH 7.0) and a high electron transfer constant (k_s) of 5.2 ± 0.6 s^{-1}. Compared to a cytochrome *c*-modified polycrystalline BDD electrode, the electron transfer was faster at the cytochrome *c*-modified nanocrystalline BDD electrode due to the incorporated sp^2 state carbon as charge transfer mediators on nanocrystalline BDD surface. It is known that amperometric biosensors based on cytochrome *c* immobilized on electrode surfaces could be used to detect H$_2$O$_2$ selectively, because cytochrome *c* as heme-containing proteins is able to electrocatalyze the reduction of H$_2$O$_2$. Investigation of the electrocatalytic activity of the cytochrome *c*-modified nanocrystalline BDD electrode toward H$_2$O$_2$ exhibited a rapid amperometric response (5 s), a wide linear range (1 - 450 μM), and a low detection limit (0.7 μM at S/N = 3). In addition, the response stability of the cytochrome *c*-modified electrode toward H$_2$O$_2$ was nearly equivalent to that of the cytochrome *c*-modified polycrystalline BDD electrode but was significantly higher than that of the cytochrome *c*-modified glassy carbon electrode.

The detection limit of the enzymatic sensors for H$_2$O$_2$ is not lower than that of H$_2$O$_2$ detection by direct electrochemical oxidation discussed in section of 4.1. However, in real biological matrices, a major problem for electrochemical detection of H$_2$O$_2$ is the coexistence of many interfering, which oxidized at a potential close to that of H$_2$O$_2$. Enzymatic sensors can be used to detect H$_2$O$_2$ selectively. Thus, the detection of H$_2$O$_2$ by nonenzymatic or enzymatic sensors has their respective advantages and disadvantages.In a word, the outstanding electrochemical properties, including wide potential window, low background current, and extreme stability, together with its inherent biocompatibility and flat surface, make the nanocrystalline BDD thin-films as an intereting candidate for the substrate of a third-generation biosensor.

6.2. Selective Detection of DA in the Presence of AA

Photochemical surface reaction between hydrogen-terminated BDD and 4-pentenoic acid was used to modify the BDD electrode surface, producing carboxyl-terminated surface (Kondo et al., 2009). The differential pulse voltammogram for the solution containing DA and AA at the above carboxyl-terminated surface showed well-separated oxidation peaks for DA and AA at 0.4 and 0.6 V, respectively, and the sensitivity enhanced distinctly than the oxygen-terminated BDD electrodes. The good selectivity and sensitivity were partly due to simple electrostatic effects and partly due to suppression of the polymerization of DA oxidation products by the terminal carboxyl groups.

7. ELECTROANALYSIS OF BDD ELECTRODES WITH DIFFERENT NANOSTRUCTURES

Despite these above advantages, the BDD electrodes do not exhibit favorable electrocatalytic activity, which limits their use to some extent. Surface modification of the BDD electrodes is necessary to improve their detection performance, however, the ideal characteristics of the BDD electrodes could be lost. Hence, it is critical to create functional BDD films without the loss of favorable properties. This can be realized by creating nanostructured BDD electrodes.

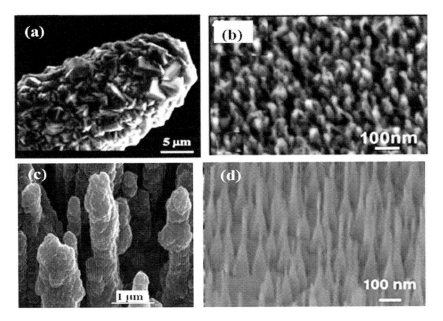

Figure 9. Scanning electron microscopy images of (a) a BDD microelectrode, (b) a BDD nanograss array, (c) a BDD nanorod forest, and (d) BDD nanowires.

7.1. In Vivo Detection at BDD Microelectrodes

The small size of the microelectrodes renders them suitable for use in vivo detection. Carbon fiber microelectrodes have been used widely for biomolecules detection in biological system because of their small size (< 30 μm), biocompatible, and temporal response (Zhang et al., 2007). However, during exposure to physiological environments, surface fouling by biomolecule adsorption makes that the response stability and sensitivity at carbon fiber microelctrodes are low. Besides, for activation of the carbon electrodes, pretreatment could introduce surface roughening, surface oxide formation, and microstructural damage. Recently, BDD microelectrodes have been fabricated by deposited a BDD film on a metal wire substrate (Park et al., 2006a,b). Scanning electron microscopy images of a micro-sized Pt wire before and after coating of BDD thin-film, and the tip of the Pt wire covered with a BDD film were showed in Figure 9a, respectively. The result of Raman spectra (not shown here) showed that the BDD microelectrode was of good quality under certain deposition condition. The BDD microelectrodes were expected to meet the above listed requirements in biological enviroments, because of its (i) hard, lubricious and biocompatible nature that enables easy penetration into tissue with minimal damage, (ii) low background current over a wide potential range, (iii) extreme high electrochemical stability ascribed to the hydrophobic sp^3-bonded carbon surface on which weak adsorption of polar biomolecules and contaminants occurs, (iv) chemical inertness. Therefore, BDD microelectrodes are attractive for in vitro electroanalytical measurements.

The catecholamines, such as, norepinephrine, epinephrine, and DA, play an important role in neurotransmission and other physiological process. For clinical and diagnostic reasons, monitoring catecholamine levels in tissue and biological fluids requires highly sensitive and reliable analytical techniques. Swain's group studied amperometric detection for norepinephrine detection in tissue or nervous system using BDD microelectrodes (Park et al., 2005, 2006a,b; Quaiserová-Mocko et al., 2008). Compared with conventional carbon fiber electrodes, the BDD microelectrodes exhibited improved response performances, including, higher sensitivity, stability, and reproducibility. Specifically, coupled with video microscopy, simultaneously amperometric monitoring with the microelectrode was used to measure norepinephrine released from rat sympathetic nerves innervating meseteric arteries and veins in vitro, and the evoked contractile response. Addtionally, by virtue of capillary zone or capillary electrophoresis, the amperometric detection was obtained very high sensitivity with a low detection limit of 51 nM (S/N = 3).

Continuous amperometry with the BDD microelectrode was used to measure serotonin overflow as an oxidation current (Patel et al., 2007; Zhao et al., 2010). With the recording microelectrode positioned about 1mm above the mucosa, serotonin released from enterochromaffin cells, was elicited by both mechnical and electrical overflow stimulation as shown in Figure 10. Some minor electrode fouling, a common problem with the oxidative detection of serotonin at the carbon fibre microelctrode, was seen for BDD microelectrode but the response was enough stable to record in vitro. The fact that the oxidation current increased in the presence of the serotonin transporter inhibitor indicated that the measured signal was associated with serotonin. Serotonin diposition in the intestinal mucosa of neonatal and adult guinea pigs was also compared using a BDD microelectrode and continuous amperometry (Bian et al., 2007). The modest change in serotonin uptake rates was also detected on the BDD microelectrode when used in combination with voltammetry (Singh et al., 2010).

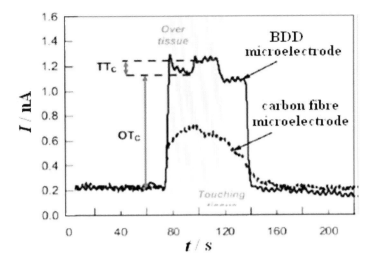

Figure 10. Continuous amperometric $i\text{-}t$ curves for 5-HT release from enterochromaffin cells of the guinea pig intestine at the BDD and carbon fibre microelectrode. The working potential was 0.7 V vs. Ag/AgCl. Krebs' buffer (pH 7.4) was flowing through the bath at 2 mL min^{-1}.

These results could help us to learn about behavioral changes and the onset of neural disease conditions (Martínez-Huitle, 2007). In view of the remarkable performance of the BDD microelectrode, it is useful for real-time measurement of samples in vivo, providing superior response sensitivity, precision, and stability as compared to a carbon fiber microelectrode.

7.2. Simultaneous Detection of DA and UA at the BDD Nanograss Array

A simple method to prepare a BDD nanograss array on a BDD film by reactive ion etching has been reported (Wei et al., 2009). The nanograss array with ~20 nm diameter and ~200 nm length has been formed on the BDD surface after oxygen plasma etching treatment, as shown in Figure 9b. The effect of the nanograss array structure on the enhancement of electrocatalytic activity of electrodes was demonstrated by detecting DA and UA. On the BDD nanograss array electrode, there are two well-defined peaks at 0.38 V and 0.78 V in neutral buffer solution, corresponding to the oxidation of DA and UA. This result is ascribed to the nanograss array on the BDD surface that could change the reactive sites, resulting in a different adsorption affinity and catalytic effect toward the different substances, thus enhancing the selectivity for some coexisting compounds. Importantly, the sensitivity has enhanced obviously compared with the oxygen-terminated BDD electrodes.

7.3. Detection of Glucose at the BDD Nanorod Forest Electrodes

Luo et al. (2009) reported that a BDD nanorod forest electrode has been fabricated by hot filament chemical vapour deposition method (Figure 9c). The BDD nanorod forest was demonstrated as a nonenzymatic glucose biosensor by simply putting it into the glucose solution. The BDD nanorod forest electrode exhibits very attractive electrochemical performance compared to conventional planar BDD electrodes, notably improved sensitivity and selectivity for glucose detection. The limit of detection was estimated at a signal-to noise ratio of 3 to 0.2 ± 0.01 µM. Selective determination of glucose in the presence of AA and UA and high stability were also obtained.

7.4. Detection of Tryptophan at the BDD Nanowires

BDD nanowires were obtained directly from the BDD film by the use of reactive ion etching with oxygen plasma (Szunerits et al., 2010). As shown in Figure 9d, the resulting nanowires are 1.4 ± 0.1 µM long with a tip and base radius of r_{tip}= 10 ± 5 nm and r_{base} = 40 ± 5 nm, respectively. The interface with the most favourable electrochemical response was investigated for the detection of tryptophan using differential pulse voltammetry. A detection limit of 0.5 µM was obtained on the BDD nanowires. This is significantly lower than on the planar BDD electrodes (section 2.6), but an order of magnitude higher than reported on glassy carbon electrodes modified with single-walled carbon nanotube films (Huang et al., 2004).

In conclusion, the nanostructured BDD electrodes can improve the reactive site, accelerate the electron transfer, promote the elelctrocataytic activity, and enhance the selectivity. They not only overcome the disadvantages of the as-grown BDD film electrodes, but also avoid the disadvantages associated with surface modification of the electrode. The above excellent properties of these nanostructured BDD electrodes indicate the promise for real biochemical applications.

CONCLUSIONS

In conclusion, several functional BDD electrodes, including hydrogen-terminated, oxygen-terminated, metal (oxide)-modified, amine-terminated, and carboxyl-terminated thin films, and BDD nanostructures, have been used for determining various bio-analytes over the past decade or so. Because of the excellent properties of BDD substrates, these BDD-based amperometric biosensors exhibited good performances in terms of high sensitivity, selectivity, reproducibilty and long-term stability. That is to say, BDD electrodes are interesting candidates for the construction of amperometric biosensors or the direct electrochemical detection of biomolecules.

Despite the impressive progress in the development of BDD-based amperometric biosensors, the promise of the application of these biosensors in real biological systems has not been fulfilled, and there are still many challenges and obstacles related to the achievement of a highly stable and reliable continuous biomolecules monitoring. Future developments will rely upon the close collaboration of analytical technology, electrochemistry, biological engineering and other relative technologies to ensure effective application and exploitation of BDD electrode materials in amperometric biosensing. There are some areas of interests opened up by these developments. One promising area is in search of reliable modification method of BDD electrode surfaces to further enhance the selectivity and sensitivity for biomolecules detection in complex systems. Another area for future advances is the BDD sensor arrays for detecting several bio-analytes at once. With the development of the electron beam and nanoimprint lithography, there is an opportunities for fabrication of ordered micro- or nanostructure on BDD thin films. Therefore, the prepared BDD microelectrode-arrays could be used for the real-time and in-vivo detection of biologically-related species in real biological systems.

REFERENCES

Bennett, J. A.; Wang, J.; Show, Y.; Swain, G. M. (2004). Effect of sp^2-Bonded nondiamond carbon impurity on the response of boron-doped polycrystalline diamond thin-film electrodes. *J. Electrochem. Soc.* 151, E306-E313.

Bian, X. C.; Patel, B.; Dai, X. L.; Galligan, J. J.; Swain, G. (2007). High mucosal serotonin availability in neonatal guinea pig ileum is associated with low serotonin transporter expression. *Gastroenterology* 132, 2438-2447.

Bouambrane, F.; Tadjeddine, A., Butler, J. E.; Tenne, R.; Lévy-Clément, C. (1996). Electrochemical study of diamond thin-films in neutral and basic solutions of nitrate. *J. Electroanal. Chem.* 405, 95-99.

Chiku, M.; Horisawa, K.; Doi, N.; Yanagawa, H.; Einaga, Y. (2010). Electrochemical detection of tyrosine derivatives and protein tyrosine kinase activity using boron-doped diamond electrodes. *Biosens. Bioelectron.* 26, 235-240.

Compton, R. G.; Foord, J. S.; Marken, F. (2003). Electrochemical detection of tyrosine derivatives and protein tyrosine kinase activity using boron-doped diamond electrodes. *Electroanalysis* 15, 1349-1363.

Febbraio, F.; Merone, L.; Cetrangolo, G. P.; Rossi, M.; Nucci, R.; Manco, G. (2011). Thermostable esterase 2 from alicyclobacillus acidocaldarius as biosensor for the detection of organophosphate pesticides. *Anal. Chem.* 83, 1530-1536.

Fischer, A. E.; Show, Y.; Swain, G. M. (2004). Electrochemical performance of diamond thin-film electrodes from different commercial sources. *Anal. Chem.* 76, 2553-2560.

Forrow, N. J.; Bayliff, S. W. (2005). Electroanalysis of dopamine and NADH at conductive diamond electrodes. *Biosens. Bioelectron.* 21, 581-587.

Fujishima, A.; Rao, T. N.; Popa, E.; Sarada, B. V.; Yagi, I.; Tryk, D. A. (1999). Electroanalysis of dopamine and NADH at conductive diamond electrodes. *J. Electroanal. Chem.* 473, 179-185.

Granger, M. C.; Witek, M.; Xu, J. S.; Wang, J.; Hupert, M.; Hanks, A.; Koppang, M. D.; Butler, J. E.; Lucazeau, G.; Mermoux, M.; Strojek, J. W.; Swain, G. M. (2000). Standard electrochemical behavior of high-quality, boron-doped polycrystalline diamond thin-film electrodes. *Anal. Chem.* 72, 3793-3804.

Haymond, S.; Babcock, G. T.; Swain, G. M. (2002). Direct Electrochemistry of cytochrome c at nanocrystalline boron-doped diamond. *J. Am. Chem. Soc.* 124, 10634-10635.

Huang, W.; Mai, G.; Liu, Y.; Yang, C.; Qua, W. (2004). Voltammetric determination of tryptophan at a single-walled carbon nanotubes modified electrode. *J. Nanosci. Nanotechnol.* 4, 423-427.

Ivandini, T. A.; Honda, K.; Rao, T. N.; Fujishima, A.; Einaga, Y. (2007). Simultaneous detection of purine and pyrimidine at highly boron-doped diamond electrodes by using liquid chromatography. *Talanta* 71, 648-655.

Ivandini, T. A.; Rao, T. N.; Fujishima, A.; Einaga, Y. (2006). Electrochemical oxidation of oxalic acid at highly boron-doped diamond electrodes. *Anal. Chem.* 78, 3467-3471.

Ivandini, T. A.; Sarada, B. V.; Rao, T. N.; Fujishima, A. (2003). Electrochemical oxidation of underivatized-nucleic acids at highly boron-doped diamond electrodes. *Analyst* 128, 924-929.

Ivandini, T. A.; Sato, R.; Makide, Y.; Fujishima, A.; Einaga, Y. (2004). Electroanalytical application of modified diamond electrodes. *Diamond Relat. Mater.* 13, 2003-2008.

Ivandini, T. A.; Sato, R.; Makide, Y.; Fujishima, A.; Einaga, Y. (2005). Pt-implanted boron-doped diamond electrodes and the application for electrochemical detection of hydrogen peroxide. *Diamond Relat. Mater.* 14, 2133-2138.

Kondo, T.; Niwano, Y.; Tamura, A.; Imai, J.; Honda, K.; Einaga, Y.; Tryk, D.A.; Fujishima, A.; Kawai, T. (2009). Enhanced electrochemical response in oxidative differential pulse voltammetry of dopamine in the presence of ascorbic acid at carboxyl-terminated boron-doped diamond electrodes. *Electrochim. Acta* 54, 2312-1319.

Koppang, M. D.; Witek, M.; Blau, J.; Swain, G. M. (1999). Electrochemical oxidation of polyamines at diamond thin-film electrodes. *Anal. Chem.* 71, 1188-1195.

Kowalczyk, A.; Nowicka, A.; Jurczakowski, R.; Fau, M.; Krolikowska, A.; Stojek, Z. (2011). Construction of DNA biosensor at glassy carbon surface modified with 4-aminoethylzenediazonium salt. *Biosens. Bioelectron.* 26, 2506-2512.

Lee, J.; Park, S.-M. (2005). Direct electrochemical assay of glucose using boron-doped diamond electrodes. *Anal. Chim. Acta* 545, 27-32.

Lin, X. H.; Zhang, Y. F.; Chen, W.; Wu, P. (2007). Electrocatalytic oxidation and determination of dopamine in the presence of ascorbic acid and uric acid at a poly (*p*-nitrobenzenazo resorcinol) modified glassy carbon electrode. *Sensor. Actuat. B* 122, 309-314.

Luo, D. B.; Wu, L. Z.; Zhi, J. F. (2009). Fabricaion of boron-doped diamond nanorod forest electrodes and their application in nonenzymatic amperometric glucose biosensing. *ACS Nano* 3, 2121-2128.

Maalouf, R.; Soldatkin, A.; Vittori, O.; Sigaud, M.; Saikali, Y.; Chebib, H.; Loir, A. S.; Garrelie, F.; Donnet, C.; Renault, N. J. (2006). Study of different carbon materials for amperometric enzyme biosensor development. *Mater. Sci. Eng. C* 26, 564-567.

Marken, F.; Paddon, C. A.; Asogan, D. (2002). Direct cytochrome c electrochemistry at boron-doped diamond electrodes. *Electrochem. Commun.* 4, 62-66.

Martínez-Huitle, C. A. (2007). Diamond microelectrodes and their applications in biological studies. *Small* 3, 1474-1476.

Nekrassova, O.; Lawrence, N. S.; Compton R. G. (2004). Direct oxidation of haemoglobin at bare boron-doped diamond electrodes. *Analyst* 129, 804-805.

Notsu, H.; Tatsuma, T.; Fujishima, A. (2002). Tyrosinase-modified boron-doped diamond electrodes for the determination of phenol derivatives. *J. Electroanal. Chem.* 523, 86-92.

Park, J.; Galligan, J. J.; Fink, G. D.; Swain, G. M. (2006). In vitro continuous amperometry with a diamond microelectrode coupled with video microscopy for simultaneously monitoring endogenous norepinephrine and its effect on the contractile response of a rat mesenteric artery. *Anal. Chem.* 78, 6756-6764.

Park, J.; Quaiserová-Mocko, V.; Pecková, K.; Galligan, J. J.; Fink, G. D.; Swain, G. M. (2006). Fabrication, characterization, and application of a diamond microelectrode for electrochemical measurement of norepinephrine release from the sympathetic nervous system. *Diamond Relat. Mater.* 15, 761-772.

Park, J.; Show, Y.; Quaiserova, V.; Galligan, J. J.; Fink, G. D.; Swain, G. M. (2005). Diamond microelectrodes for use in biological environments. *J. Electroanal. Chem.* 583, 56-58.

Patel, B. A.; Bian, X. C.; Quaiserová-Mocko, V.; Galligan, J. J.; Swain, G. M. (2007). In vitro continuous amperometric monitoring of 5-hydroxytryptamine release from enterochromaffin cells of the guinea pig ileum. *Analyst* 132, 41-47.

Perret, A.; Haenni, W.; Skinner, N.; Tang, X.-M.; Gandini, D.; Comninellis, C.; Correa, B.; Foti, G. (1999). Electrochemical behavior of synthetic diamond thin film electrodes. *Diamond Relat. Mater.* 8, 820-823.

Popa, E.; Kubota, Y.; Tryk, D. A.; Fujishima, A. (2000). Selective voltammetric and amperometric detection of uric acid with oxidized diamond film electrodes. *Anal. Chem.* 72, 1724-1727.

Popa, E.; Notsu, H.; Miwa, T.; Tryk, D. A.; Fujishima, A. (1999). Selective electrochemical detection of dopamine in the presence of ascorbic acid at anodized diamond thin film electrodes. *Electrochem. Solid-State Lett.* 2, 49-51.

Prado, C.; Flechsig, G.-U.; Gründler, P.; Foord, J. S.; Marken, F.; Compton, R. G. (2002). Electrochemical analysis of nucleic acids at boron-doped diamond electrodes. *Analyst* 127, 329-332.

Quaiserová-Mocko, V.; Novotný, M.; Schaefer, L. S.; Fink, G. D.; Swain, G. M. (2008). CE coupled with amperometric detection using a boron-doped diamond microelectrode: Validation of a method for endogenous norepinephrine analysis in tissue. *Electrophoresis* 29, 441-447.

Rajesh; Takashima, W.; Kaneto, K. (2004). Amperometric phenol biosensor based on covalent immobilization of tyrosinase onto an electrochemically prepared novel copolymer poly (*N*-3-aminopropyl pyrrole-co-pyrrole) film. *Sens. Actuat. B* 102, 271-277.

Rao, T. N.; Yagi, I.; Miwa, T.; Tryk, D. A.; Fujishima, A. (1999). Electrochemical oxidation of NADH at highly boron-doped diamond electrodes. *Anal. Chem.* 71, 2506-2511.

Sanz, V. C.; Mena, M. L.; González-Cortés, A.; Yáñez-Sedeño, P.; Pingarrón, J. M. (2005). Development of a tyrosinase biosensor based on gold nanoparticles-modified glassy carbon electrodes: Application to the measurement of a bioelectrochemical polyphenols index in wines. *Anal. Chim. Acta.* 528, 1-8.

Sarada, B. V.; Rao, T. N.; Tryk, D. A.; Fujishima, A. (2000). Electrochemical oxidation of histamine and serotonin at highly boron-doped diamond electrodes. *Anal. Chem.* 72, 1632-1638.

Singh, Y. S.; Sawarynski, L. E.; Michael, H. M.; Ferrell, R. E.; Murphey-Corb, M. A.; Swain, G. M.; Patel, B. A.; Andrews, A. M. (2010). Boron-doped diamond microelectrodes reveal reduced serotonin uptake rates in lymphocytes from adult rhesus monkeys carrying the short allele of the 5-HTTLPR. *ACS Chem. Neurosci.* 1, 49-64.

Spătaru, N.; Sarada, B. V.; Popa, E.; Tryk, D. A.; Fujishima, A. (2001). Voltammetric determination of L-cysteine at conductive diamond electrodes. *Anal. Chem.* 73, 514-519.

Szunerits, S.; Coffinier, Y.; Galopin, E.; Brenner, J.; Boukherroub, R. (2010). Preparation of boron-doped diamond nanowires and their application for sensitive electrochemical detection of tryptophan. *Electrochem. Commun.* 12, 438-441.

Terashima, C.; Rao, T. N.; Sarada, B. V.; Fujishima, A. (2003). Amperometric detection of oxidized and reduced glutathione at anodically pretreated diamond electrodes. *Chem. Lett.* 32, 136-137.

Terashima, C.; Rao, T. N.; Sarada, B. V.; Spataru, N.; Fujishima, A. (2003). Electrodeposition of hydrous iridium oxide on conductive diamond electrodes for catalytic sensor applications. *J. Electroanal. Chem.* 544, 65-74.

Toghill, K.; Xiao, L.; Phillips, M. A.; Compton, R. G. (2010). The non-enzymatic determination of glucose using an electrolytically fabricated nickel microparticle modified boron-doped diamond electrode or nickel foil electrode. *Sensor. Actuat. B* 147, 642-652.

Tryk, D. A.; Tachibana, H.; Inoue, H.; Fujishima, A. (2007). Boron-doped diamond electrodes: The role of surface termination in the oxidation of dopamine and ascorbic acid. *Diamond Relat. Mater.* 16, 881-887.

Wang, B.; Zhang, J.; Dong, S. (2000). Silica sol–gel composite film as an encapsulation matrix for the construction of an amperometric tyrosinase-based biosensor. *Biosens. Bioelectron.* 15, 397-402.

Wang, J. (2008). Electrochemical glucose biosensors. *Chem. Rev.* 108, 814-825.

Wang, J.; Chen, G.; Muck, A.; Jr.; Shin, D.; Fujishima, A. (2004). Microchip capillary electrophoresis with a boron-doped diamond electrode for rapid separation and detection of purines. *J. Chromatography A* 1022, 207-212.

Wang, J.; Lin, Y. H. (2008). Functionalized carbon nanotubes and nanofibers for biosensing applications. *TrAC Trends Anal. Chem.* 27, 619-626.

Wei, M.; Sun, L.-G.; Xie, Z.-Y.; Zhi, J.-F.; Fujishima, A.; Einaga, Y.; Fu, D.-G.; Wang, X.-M.; Gu, Z.-Z. (2008). Selective determination of dopamine on a boron-doped diamond electrode modified with gold nanoparticle/polyelectrolyte-coated polystyrene colloids. *Adv. Funct. Mater.* 18, 1414-1421.

Wei, M.; Terashima, C.; Lv, M.; Fujishima, A.; Gu, Z. Z. (2009). Boron-doped diamond nanograss array for electrochemical sensors. *Chem. Commun.* 24, 3624-3626.

Weng, J.; Xue, J. M.; Wang, J.; Ye, J.-S.; Cui, H. F.; Sheu, F.-S.; Zhang, Q. Q. (2005). Gold-cluster sensors formed electrochemically at boron-doped-diamond electrodes: detection of dopamine in the presence of ascorbic acid and thiols. *Adv. Funct. Mater.* 15, 639-647.

Wilson, G. S.; Hu, Y. B. (2000). Enzyme-based biosensors for in vivo measurements. *Chem. Rev.* 100, 2693-2704.

Yang, W.; Auciello, O.; Butler, J. E.; Cai, W.; Carlisle, J. A.; Gerbi, J. E.; Gruen, D. M.; Knickerbocker, T.; Lasseter, T. L.; Russell, J. N.; JR.; Smith, L. M.; Hamers, R. J. (2002). DNA-modified nanocrystalline diamond thinfilms as stable, biologically active substrates. *Nature Mater.* 1, 253-257.

Yang, W.; Baker, S. E.; Butler, J. E.; Lee, C.; Russell, J. N.; Jr., Shang, L.; Sun, B.; Hamers, R. J. (2005). Electrically addressable biomolecular functionalization of conductive nanocrystalline diamond thin films. *Chem. Mater.* 17, 938-940.

Zhang, L.; Jiang, X. (2005). Attachment of gold nanoparticles to glassy carbon electrode and its application for the voltammetric resolution of ascorbic acid and dopamine. *J. Electroanal. Chem.* 583, 292-299.

Zhang, M. N.; Gong, K. P.; Zhang, H. W.; Mao, L. Q. (2005). Layer-by-layer assembled carbon nanotubes for selective determination of dopamine in the presence of ascorbic acid. *Biosens. Bioelectron.* 20, 1270-1276.

Zhang, M. N.; Liu, K.; Xiang, L.; Lin, Y. Q.; Su, L.; Mao, L. Q. (2007). Carbon nanotube-modified carbon fiber microelectrodes for in vivo voltammetric measurement of ascorbic acid in rat brain. *Anal. Chem.* 79, 6559-6565.

Zhang, Y.; Zeng, G.-M.; Tang, L.; Huang, D.-L.; Jiang, X.-Y.; Chen, Y.-N.; (2007). A hydroquinone biosensor using modified core-shell magnetic nanoparticles supported on carbon paster electrode. *Biosens. Bioelectron.* 22, 2121-2126.

Zhao, G.H.; Qi, Y.; Tian, Y. (2006). Simultaneous and direct determination of tryptophan and tyrosine at boron-doped diamond electrode. *Electroanalysis* 18, 830-834.

Zhao, H.; Bian, X.C.; Galligan, J. J.; Swain, G.M. (2010). Electrochemical measurements of serotonin (5-HT) release from the guinea pig mucosa using continuous amperometry with a boron-doped diamond microelectrode. *Diamond Relat. Mater.* 19, 182-185.

Zhao, J. W.; Wu, D. H.; Zhi, J. F. (2009a). A direct electrochemical method for diabetes diagnosis based on as-prepared boron-doped nanocrystalline diamond thin-film electrodes. *J. Electroanal. Chem.* 626, 98-102.

Zhao, J. W.; Wu, L. Z.; Zhi, J. F. (2009b). Non-enzymatic glucose detection using as-prepared boron-doped diamond thin-film electrodes. *Analyst* 134, 794-799.

Zhou, Y. L.; Tian, R. H.; Zhi, J. F. (2007). Amperometric biosensor based on tyrosinase immobilized on a boron-doped diamond electrode. *Biosens. Bioelectron.* 22, 822-828.

Zhou, Y. L.; Zhi, J. F. (2006). Development of an amperometric biosensor based on covalent immobilization of tyrosinase on a boron-doped diamond electrode. *Electrochem. Commun.* 8, 1811-1816.

Zhou, Y. L.; Zhi, J. F.; Zou, Y. S.; Zhang, W. J.; Lee, S. T. (2008). Direct electrochemistry and electrocatalytic activity of cytochrome *c* covalently immobilized on a boron-doped nanocrystalline diamond electrode. *Anal. Chem.* 80, 4141-4146.

Chapter 2

OPTICAL PROPERTIES OF DIAMOND LIKE CARBON THIN FILMS

Sk. Faruque Ahmed[*]

*Department of Physics, Aliah University, DN - 41, Sector - V,
Salt Lake, Kolkata 700 091, India*

ABSTRACT

Carbon-based materials, clusters, and molecules are unique in many ways. Carbon based materials are ideally suitable as molecular level building blocks for nanoscale systems design, fabrication and applications. From a structural or functional materials perspective, carbon is the only element that exists in a variety of shapes and forms with varying physical and chemical properties. One distinction relates to the many possible configurations of the electronic states of a carbon atom, which is known as the hybridization of atomic orbitals and relates to the bonding of a carbon atom to its nearest neighbors.

Diamond like carbon (DLC) can be defined as a metastable phase of amorphous carbon (a-C) or hydrogenated amorphous carbon (a-C:H) containing a significant fraction of sp^3 hybridization. As well, it is commonly accepted that the graphitic sp^2 clusters are embedded in the amorphous sp^3 bonded matrix, where the collective behavior of sp^2 sites is responsible for the optical and electrical properties, whereas the sp^3 regions govern mechanical properties. DLC has some extreme properties like diamond, such as high mechanical hardness, chemical inertness, dielectric strength, low wear and friction, optical transparency in the visible and infrared region, high electrical resistivity, high thermal conductivity and low electron affinity. These properties are promising for a wide range of applications, as protective coatings in areas, such as optical windows, magnetic storage disks, car parts, biomedical applications and as micro-electromechanical devices and electronic applications.

In this chapter, the synthesis and characterizations particularly optical properties of the different metal and nonmetal doped DLC thin films have been discussed in details.

---

[*] Corresponding author: faruquekist@gmail.com.

1. INTRODUCTION

Carbon is the sixth element of the periodic table and has the lowest atomic number of any element in column IV of the periodic table. In contrast to Si, Ge and Sn, which have the same number of electrons in the outermost shell as carbon and can only exist in cubic sp^3 hybridization, carbon not only exhibits sp^3 hybridization (diamond) but also planar sp^2 hybridization as in the graphite structure and sp^1 hybridization as in carbynes.

Carbon based materials attract a growing interest due to their specific characteristics such as: (I) resistance to acid / basic media, (II) possibility to control up to certain limits, the porosity and surface chemistry and (III) easy recovery of precious metals by support burning resulting in a low environmental impact.

Diamond like carbon (DLC) or amorphous hydrogenated carbon (a:C-H) is the generic name of a mixture of graphite and diamond-type bound carbon. Also, it is well known that in DLC graphitic sp^2 clusters are embedded in the amorphous sp^3 bonded matrix, where the collective behavior of sp^2 sites is responsible for the optical and electrical properties, whereas the sp^3 regions govern mechanical properties. The DLC thin film has been extensively studied all over the world during the last three decades, as it possesses several novel properties including high mechanical hardness, dielectric strength, chemical inertness, low wear and friction, optical transparency in the visible and infrared region, high electrical resistivity, high thermal conductivity and low electron affinity. These properties are promising for a wide range of applications, such as in mechanical, optical, electronic and medical fields. Instead of these interesting properties, DLC films have several inherent problems, such as their high internal stress and low thermal stability. The high intrinsic stress generated in the film causes some serious problems in various applications of this material, like degradation of electrical and optical properties and leads to delamination of the films from the substrate and poor adhesion.

The ratio of sp^2/sp^3 carbon atoms is one of the most important factors determining the quality of the DLC films, which can be changed by incorporating different metal or non-metal into DLC matrix. Metal and nonmetal doped DLC thin films have been an interesting field since last few years because, they have a great potential for solving some of the major drawbacks of pure DLC thin films. Various attempts have been made to dope DLC thin films with different nonmetals and metals such as nitrogen, iodine, boron, phosphorous, fluorine, gold, silver, titanium, tin, cupper etc. and the doping effects of these elements have been extensively investigated in terms of the chemical, electrical, optical, and mechanical properties of DLC films. Doping of different elements in the DLC thin films reduced residual internal stress without sacrificing the hardness, leads to good adhesion to metal alloys, steels and glasses, improved high temperature stability, reduced hydrogen loss, graphitization and resistance to oxidation. Due to these noble properties of doped DLC thin film, it has been studied in the last few years and has potential applications in many areas, such as electronic applications, solar cells, photoluminescence cells, optoelectronic devices, field effect transistor and high-temperature engineering materials.

2. CHEMICAL BONDING OF CARBON

Carbon is the most versatile element in the periodic table, owing to the type, strength, and number of bonds it can form with many different elements. The diversity of bonds and their corresponding geometries enable the existence of structural isomers and geometric isomers. These are found in large, complex, and diverse structures and allow for an endless variety of organic molecules. In the periodic table, which is an arrangement of chemical elements exhibiting certain regular periodic occurrences in their behavior, the element carbon occupies the sixth position and has a molar mass 12.011 gmol^{-1}. Since 1961, all relative atomic masses have been based on a scale where the atomic mass of ^{12}C is exactly 12.00000. From a structural or functional materials perspective, carbon is the only element that exists in a variety of shapes and forms with varying physical and chemical properties.

The number of electrons in an atom is equal to the atomic number, which for carbon is 6 and the electronic ground state for carbon has the configuration $1s^2$, $2s^2$, $2p^2$. Of the six electrons in the neutral atom, four are available for the formation of chemical bonds in the outer L shell. Orbitals characterizing an atom differ from each other in terms of r, the distance of the electron from the nucleus. The electrons can be considered as a charge cloud with varying density and determined by the wave function.

The single $1s$ electron of hydrogen has its most probable location within a sphere whose radius is equal to the radius of orbit of the electron with the nucleus at the center and has no directional characteristics. With p, d and f orbitals, the situation is more complex and they are found in sets of 3, 5 and 7. It is impossible to determine the direction of any one orbital in a given set. However, the axes along which a given set lies are at definite angles to each other in space, with the nucleus at their intersection. Hence p orbitals with the nucleus at the center are at mutually right angles and lie along three cartesian co-ordinates p_x, p_y, p_z, emphasizing the directional character. An orbital, such as a p type, is dumb-bell shaped and has two parts, each situated on either side of a node, which is a region where the probability of finding an electron would very remote (Figure 2.1). A simple picture of an electron orbital is to consider the orbital as an electron cloud with the densest region representing the vicinity where the possibility of finding the electron is greatest.

One distinction relates to the many possible configurations of the electronic states of a carbon atom, which is known as the hybridization of atomic orbitals and relates to the bonding of a carbon atom to its nearest neighbors.

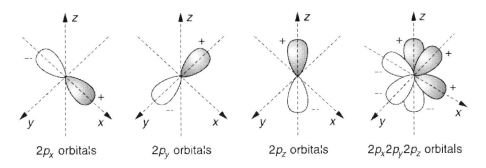

Figure 2.1. Probable location of electrons in $2p$ orbitals.

Each carbon atom has six electrons which occupy $1s^2$, $2s^2$, and $2p^2$ atomic orbitals. The $1s^2$ orbital contains two strongly bound core electrons. Four more weakly bound electrons occupy the $2s^2 2p^2$ valence orbitals. In the crystalline phase, the valence electrons give rise to $2s$, $2p_x$, $2p_y$, and $2p_z$ orbitals which are important in forming covalent bonds in carbon materials. Since the energy difference between the upper $2p$ energy levels and the lower $2s$ level in carbon is small compared with the binding energy of the chemical bonds, the electronic wave functions for these four electrons can readily mix with each other, thereby changing the occupation of the $2s$ and three $2p$ atomic orbitals so as to enhance the binding energy of the carbon atom with its neighboring atoms. The general mixing of $2s$ and $2p$ atomic orbitals is called hybridization, whereas the mixing of a single $2s$ electron with one, two, or three $2p$ electrons is called sp^n hybridization with $n = 1, 2, 3$.

If the p states (p_x, p_y, p_z) of the neighbouring ions mix with their s states, then there are four bonding and four antibonding σ orbitals and the so-called sp^3 hybridization. sp^3 hybridization leads to a tetrahedral local structure, in which the reference carbon ion is placed in the centre of the tetrahedron (at a mean angle of $109°28'$ to one another) and the four bonds connect it with the vertexes of the tetrahedron where the neighbouring ions sit (Figure 2.2(a)). The resulting structure is tridimensional, and by the repetition of this block the diamond structure is built. The distance between neighbouring atoms in diamond is 0.154 nm.

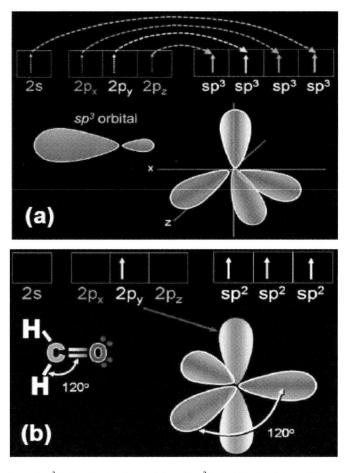

Figure 2.2. Formation of sp^3 hybridized orbitals (a) and sp^2 hybridized orbitals (b).

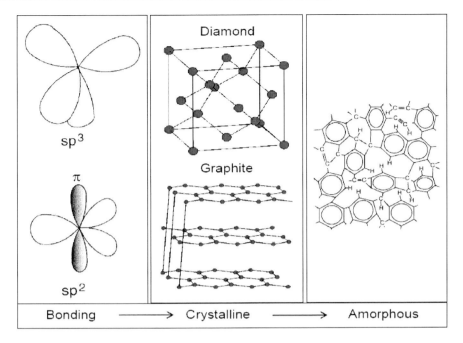

Figure 2.3. Schematic representation of different types of hybridized bonding and the resultant crystalline and model amorphous structure.

If the *s* states mix with two corresponding *p* states for each atom, then there are three bonding and three antibonding σ orbitals. The remaining *p* states hybridize to form a π state. This is the *sp²* hybridization and the *sp²* hybridization leads to three σ-bonds lying in the same plane and forming 120° angles between any two of them (Figure 2.2(b)). The repetition of this block leads to a bidimensional planar structure based on hexagonal blocks, the graphene sheet. The distance between neighbouring atoms is 0.142 nm, shorter than in diamond. This is due to the presence of π orbitals that strengthen the bond between neighbouring ions. In-plane bonding is hence determined by a concurrence of σ and π orbitals. The out-of plane bonding between graphene sheets in graphite is due to weak Vander Waals forces, and this justifies the prompt delamination of graphite. Finally, if the *s* states mix with one *p* state for each neighbouring atom, then there are two bonding and two antibonding σ orbitals and two additional π states. This is the *sp¹* hybridization and is found, for instance, in acetylene.

The σ and π orbitals have very differently shaped electron clouds. σ orbitals have the highest density of the electron cloud along the axis connecting the two neighbouring nuclei and form a strong, almost always covalent, bond between the ions. Each π orbital has its electron cloud formed by two lobes symmetric with respect to a plane containing the axis connecting the two nuclei and absent on such plane. The π-bonds formed between two nuclei are generally quite weak. Hence the energy spacing between π-bonding and antibonding states is much lower than that between σ-bonding and antibonding states. Although this is the situation for a pair of carbon ions, things can be quite heavily modified if the collection of ions is enlarged. In such a case, several additional effects come into play due to the delocalization of π orbitals. For an example, in a benzene ring, all π orbitals arising form the six carbon ions have merged to form a delocalized orbital containing all π electrons. This orbital now forms a strong additional bond between all atoms of the ring. It becomes very costly in term of energy to break a single π-bond, as it is fused with the others so that the structure is overall much more stable. π orbitals have other peculiar properties that are of relevance in organic chemistry and in life itself. For instance, the resonance plays a relevant role in determining the properties of aromatic and aliphatic structures. These bond hybridizations are not restricted to the crystalline forms of carbon and are available to form amorphous carbon films (Figure 2.3).

3. DIAMOND LIKE CARBON (DLC) AS A MATERIAL

Diamond like carbon (DLC) is often discussed together with crystalline diamond films, even though their structure is significantly different. This is because the diamond like solids exhibit unusual hardness and other properties resembling that of diamond and they are sometimes made with crystalline diamond in the same process. The diamond like solids have very high number densities of atoms compared with other conventional hydrocarbons and hydrocarbon polymers. This high number density arises from their completely cross-linked structure and provides another point of similarity with crystalline diamond.

DLC can be defined as a metastable phase of amorphous carbon (a-C) or hydrogenated amorphous carbon (a-C:H) containing a significant fraction of sp^3 hybridization. Also, it is well known that in DLC is the generic name of a mixture of graphite and diamond-type bound carbon [1]. The composition of DLC is most conveniently displayed on a ternary phase diagram as in Figure 3.1, first used by Jacob and Moller [2].

If the fraction of sp^3 bonding reaches a high degree, McKenzie [3] suggested to denote the a-C as tetrahedral amorphous carbon (ta-C), to distinguish it from sp^2 a-C. Koidl et al [4] produced DLC by plasma enhanced chemical vapour deposition (PECVD) method; whose composition reached the interior of the triangle, this was named as a-C:H. A more sp^3 bonded material with less hydrogen content, was denoted as hydrogenated tetrahedral amorphous carbon (ta-C:H) by Weiler et al [5]. Typical properties of the various forms of a-C are compared to diamond and graphite in Table - 3.1.

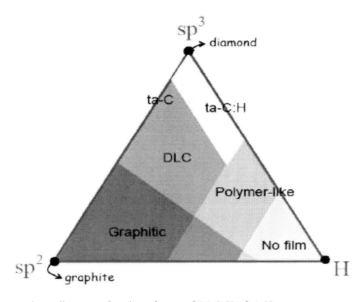

Figure 3.1. Ternary phase diagram of various forms of DLC [Ref. 1,2].

Table 3.1. Comparison of major properties of a-C with those of reference materials like diamond, graphite and C$_{60}$

Sample	sp^3 (%)	H (%)	Density (gm-cm^{-3})	Optical Gap (eV)	Hardness (Gpa)	Reference
Diamond	100	0	3.515	5.5	100	6
Graphite	0	0	2.267	0	-	7
C$_{60}$	0	0	-	1.6	-	8
Glassy C	0	0	1.3-1.55	0.01	3	9
Evaporated C	0	0	1.9	0.4-0.7	3	9
Sputtered C	5	0	2.2	0.5	-	10
ta-C	80-88	0	3.1	2.5	80	8,11
a-C:H hard	40	30-40	1.6-2.2	1.1-1.7	10-20	12
a-C:H soft	60	40-50	1.2-1.6	1.7-4.0	<10	12
ta-C:H	70	30	2.4	2.0-2.5	50	13,14

The structure of DLC is still a matter of research. However, some overall feature of this solid has become clear in the past several years. A proposed structure of DLC is shown in Figure 3.2. The shaded circles are sp^3 coordinated carbon, tetrahedrally bonded to four other atoms. The fully black circles represent sp^2 carbon atoms, trigonally bonded to three other atoms. The open circles are hydrogen atoms. One major feature of DLC is the type of clustering that the sp^2 carbon atoms undergo. It is proved, on the basis of optical spectroscopic evidences that, many of the sp^2 carbon atoms are present as large graphite-like clusters within DLC. These clusters appear to be spatially isolated from one another, because the films have extremely low conductivity.

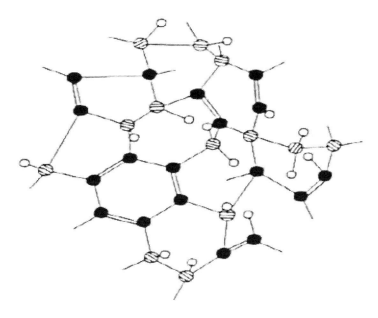

Figure 3.2. A two-dimensional representation of a proposed structure for DLC [Ref. 1].

The sp^2 hybridized carbon atoms tend to exist as nm-sized clusters in the form of graphitic rings giving rise to π and π^* bands, whereas the sp^3 component tends to be in the form of aliphatic chains. In the sp^3 configuration, as in diamond, each of the four valance electrons of a carbon atom is assigned to a tetrahedrally directed sp^3 orbital, which makes a strong σ bond to an adjacent atom. In the three-fold coordinated sp^2 configuration, as in graphite, three of the four valance electrons enter trigonally directed sp^2 orbital, which form σ bonds in a plane. The fourth valance electron lies in a p π orbital, which lies normal to the σ bonding plane. In the sp^1 configuration, two of the four valance electrons form σ bonds directed along the $\pm$ x axis, and the other two electrons enter p π orbitals in the y and z directions. The optoelectronic properties of these films are usually discussed in terms of transitions between occupied π states to unoccupied π^* states with the σ and σ^* states (associated with sp^3 C) further separated from each other.

The simplest model of the atomic network of DLC [15] is based on the behaviour of amorphous hydrogenated silicon (a-Si:H), where the bonding tends to follow the principle of local chemical equilibrium. It was assumed that in a network of amorphous hydrogenated carbon (a-C:H), the sp^3 and sp^2 sites arrange themselves to minimize the total energy. Analysis of this condition results in the 'cluster model' of DLC [15]. In this model, the non-local π bonding energy encourages sp^2 sites to pair up to give C=C bonds, as in ethylene, then combine to form planar 6-fold benzene rings, and finally to cluster together to form aromatic, graphitic clusters, within a sp^3 bonded matrix. This clustering process maximizes the π bonding energy. The σ and σ^* states then form the deep valance and conduction states, while the π and π^* states form the band edge states, thereby determining the bandgap, as shown schematically in Figure 3.3.

This model was supported by the realization that the surprisingly low optical gap of DLC could be explained in terms of graphitic clusters [15], with the band gap E_g varying with the number of rings in a cluster as $E_g = 6/M^{1/2}$ eV, where M is the number of six-fold rings in the cluster.

In most amorphous semiconductors, the Tauc band gap represents the transitions between extended states and in the case of a-C:H, in the absence of disorder, the Tauc gap is given by the separation between these two bands $2E_\pi$. The larger the average sizes of the sp^2 cluster the smaller the Tauc gap. An alternative band gap that is sometimes employed is to measure the energy at which the absorption coefficient is 10^4 cm^{-1}; this is known as the E_{04} gap.

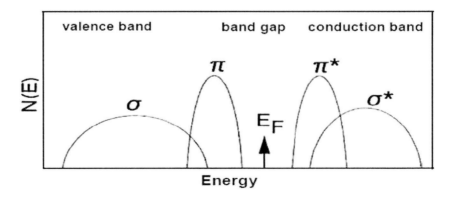

Figure 3.3. Schematic diagram of σ and states in amorphous carbons (a-C) [Ref. 15].

It is now well established that in the case of metals the conductivity is determined by diffusive electronic states. But in the case of amorphous materials, disorder plays an important role with a net reduction of the mobility of the carriers via scattering. Anderson showed that in a sufficiently disordered one-band model no movement of carriers can take place without thermal activation [16]. Tunneling between localised states becomes possible at higher temperatures and the extent of electron hopping is controlled by the properties of the mid-gap and tail states.

Studies of DLC have been performed extensively since 1971, when Aisenberg and Chabot [17] first prepared such films, and there has been explosive attention in understanding the growth mechanisms, material properties and usage in industrial applications. DLC has been extensively studied all over the world during the last few decades, as it possesses properties close to those of diamond, such as high mechanical hardness, chemical inertness, dielectric strength, low wear and friction, optical transparency in the visible and infrared region, high electrical resistivity, high thermal conductivity and low electron affinity which make them a powerful candidate for the next generation of high performance electronic devices and other vacuum micro-electronic devices [4,9,18-21]. At present, high quality diamond films can be produced at high temperatures (>500 °C) that tend to develop preferential growth of individual grains, thus resulting in faceted and rough film surfaces. On the other hand, DLC films can be deposited at low temperature on a large area substrate with high deposition rates; hence the range of materials it can coat is greatly increased and low-cost, large area electronic devices can be easily realized. Also, there are some practical advantages of DLC over diamond. DLC can be smooth on nanometer scale, with its hardness approaching that of diamond. DLC films have widespread applications. The comparison of some major property of DLC over diamond and graphite are given in Table 3.2.

DLC films have widespread applications, as protective coatings in areas, such as optical windows, magnetic storage disks, car parts, biomedical applications, photovoltaic cell, field emission display and as micromechanical devices (MEMs) [22-28]. One of the most interesting properties of DLC is the ability to vary the optical band gap by simply varying the relative proportion of the sp^2/sp^3 bond hybridization. This allows for the formation of band gap modulated super-lattice structures based on a single homogeneous material system [29]. Typical properties of DLC and its different applications are shown in Table 3.3.

Table 3.2. Comparison of some major property of DLC over diamond and graphite

Property	Diamond	DLC	Graphite
Crystal structure	Cubic, a = 3.57 Å; Hexagonal, a = 2.52 Å, c = 4.12 Å	Amorphous with small crystal regions mixed with sp^2 and sp^3 bonds	Hexagonal a = 2.47 Å, c = 6.71 Å
Fraction of hybrid orbital	100 % sp^3	40-75 % sp^3 20-60 % sp^2; ~2 % sp^1	100 % sp^2
Density (g/cm^3)	3.51	1.6 – 2.0	2.26
Hardness (GPa)	70 - 100	20 – 80	< 5
Friction Coefficient	0.05	0.03 – 0.2	0.1
Refractive Index	2.42	1.8 – 2.6	2.15 – 1.8
Transparency	UV-VIS-IR	VIS-IR	Opaque
Optical gap (eV)	5.45	1.0 – 2.6	0
Resistivity (Ω-cm)	> 10^{16}	$10^2 – 10^{14}$	0.2 – 0.4

Table 3.3. Summary of properties and applications of DLC films

Property	Type of use	Applications
Transparency in Vis and IR; optical band gap = 1.0-4.0 eV	Optical coatings	Antireflective and wear-resistant coatings for IR optics
Chemical inertness to acids, alkalis and organic solvents	Chemically passivating coatings	Corrosion protection of magnetic media, biomedical
High hardness; 5-80 Gpa; low friction coefficient < 0.01- 0.7	Tribological, wear-resistant coatings	Magnetic hard drives, magnetic tapes, razor blades, bearings, gears
Nanosmooth	Very thin coatings < 5 nm	Magnetic media
Wide range of electrical resistivities = 10^2-10^{16} Ω–cm	Insulating coatings	Insulating films
Low dielectric constants < 4	Low-k dielectrics Field emission	Interconnect dielectrics, Field emission flat panel displays

Although due to extensive research activities in the field of diamond and DLC during the last three decades, the materials and the technology are more or less well understood by the time, still there are numerous problems with these materials, which need motivated works. For example, the high intrinsic stress generated in the film causes some serious problems in various applications of this material, like degradation of electrical and optical properties and leads to peeling-off of the films from the substrate and poor adhesion.

By changing co-ordination number of the carbon atoms in the DLC, it is possible to reduce the generated stress. This can be achieved by introducing foreign atoms in the DLC matrix. Doping of DLC with n and p-type dopants such as nitrogen, phosphorous, boron, iodine, fluorine [30-34] etc. and their doping effects on the electrical, optical and mechanical properties of DLC films have been extensively investigated. Boron acts as a good acceptor and is well incorporated into the structure of DLC analogous with the case in diamond. In nitrogen doped DLC films, the optical band gap and the fraction of the sp^3 bonding were both found to remain constant, they only decreased at higher N contents. Kundoo et al reported the optical and electrical properties of DLC films drastically change by boron and nitrogen doping [35]. Room temperature electrical conductivity increased by two orders of magnitude in N-DLC films and one order of magnitude in B-DLC films respectively. Incorporation of nitrogen and boron reduced the electrical activation energy of the films from 0.75 eV for pure DLC to 0.32 eV and 0.58 eV respectively for doped films. Optical measurements in UV-Visible-NIR region showed an increase in the optical band gap energy from 1.0 eV to 2.12 by N doping where as by B doping it increased up to 2.0 eV. The films showed high transparency in visible region [35]. However, the doping efficiency is very low. It has been claimed that the decrease of resistivity by incorporation of dopants may be related to a dopant induced graphitization [36]. The electronic transport mechanism was investigated in pulsed laser deposited ta-C by correlating the effect of annealing on electrical conductivity and stress relaxation and using the latter to estimate the changes in sp^2 carbon concentrations [37]. The results indicated that, in the specific films, the conductivity is an exponential function of the changes in the sp^2 concentration, indicating a tunneling or hopping transport mechanism. The electronic transport could be modeled as taking place by thermally activated conduction along linkages or chains of sp^2 carbon atoms with variable range and variable orientation hopping, resulting in heterogeneity of conduction pathways through the sample.

Metal incorporated DLC (Me-DLC) films have aroused immense interest among the workers due to their interesting tribological, electrical and optical properties. Nanometer-sized clusters incorporated into the DLC matrix can drastically change different properties for the application point of view. Various groups have tried to deposit Me-DLC using different metal elements such as Au, Ta, Ti, Li, Sn, Mo, Al etc. [38-42]. Czyzniewski et al [39] reported the deposition of tungsten incorporated DLC (W-DLC) nanocomposite coatings on steel substrates below 423 K temperature and discussed their mechanical properties. Rusli et al [40] have shown that resistivities of their tungsten incorporated a-C films (W-C:H) decreased drastically upon incorporation of several atomic percentage of tungsten. Schuler et al [41] have studied the optical properties in details of the Ti-C:H carbon films deposited by reactive magnetron sputtering. Huang et al [42] have studied in detail the dielectric properties of Mo-C:H films deposited by ECRCVD process. Due to incorporation of nanocrystalline tin (Sn) particles into the amorphous DLC matrix, the resistivity of the films decreased drastically compared to the undoped DLC film and the threshold field for electron emission decreased to 3.5 V/µm for an optimum Sn % from 15 V/µm for the undoped DLC film [43].

4. BASIC THEORY OF OPTICAL PROPERTIES FOR THIN FILM

When a ray of light is incident normally at an interface between two media with refractive indices n_0 and n_1 (which are complex in most general cases), multiple reflections take place inside the interface and the reflectance R and transmittance T (defined as ratios of reflected and transmitted energy to the incident energy) are given by [44]

$$R = \frac{(n_0 - n_1)(n_0 - n_1)*}{(n_0 + n_1)(n_0 + n_1)*} \qquad (1)$$

$$T = \frac{(2n_0 n_1)(2n_0 n_1)*}{(n_0 + n_1)(n_0 + n_1)*} \qquad (2)$$

where the asterix ($*$) denotes the complex conjugate. [$(n_0-n_1)/(n_0+n_1)$] and [$(2n_0n_1)/(n_0+n_1)$] are called refelectivity r and transmittivity t, respectively, or Fresnel coefficients. For angles of incidence other than normal, r and t depend on the polarization of the incident wave. For a wave with the electric vector in the plane of incidence (TM mode) Fresnel reflection coefficient can be written as

$$r_p = \frac{n_0 \cos\theta_1 - n_1 \cos\theta_0}{n_0 \cos\theta_1 + n_1 \cos\theta_0} \qquad (3)$$

where θ_0 and θ_1 are angles of incidence and refraction respectively. For a thin film on a substrate (as shown in Figure 4.1) we have to consider two interfaces: the incident medium (refractive index, n_0)/film (refractive index, n) – interface 1 and film/substrate (refractive index, n_1) – interface 2.

Taking into account the multiple reflections at the two interfaces the amplitude for reflected and transmitted waves can be written as [44]

$$A_r = \frac{r_1 + r_2 e^{-2i\delta_1}}{1 + r_1 r_2 e^{-2i\delta_1}} \tag{4}$$

$$A_t = \frac{t_1 t_2 e^{-i\delta_1}}{1 + r_1 r_2 e^{-2i\delta_1}} \tag{5}$$

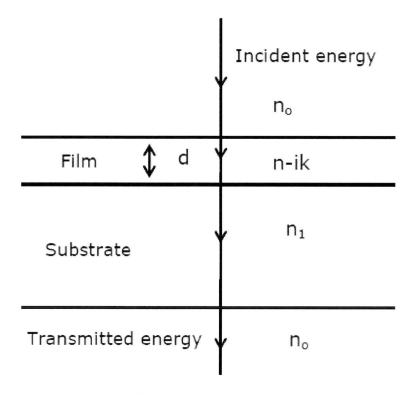

Figure 4.1. System of absorbing thin film on a thick finite transparent substrate.

where r_1, r_2 and t_1, t_2 are reflection and transmission coefficients at the front and rear faces of the film. δ is the change in phase of the beam on traversing the film and is given by

$$\delta = \frac{2\pi}{\lambda} nd \cos\theta_0 \tag{6}$$

in which d is the thickness of the film and θ_0 is the angle of incidence in the first medium. Assuming unit amplitude of the incident wave in the first medium, the reflectance and transmittance can be written as

$$R = \frac{r_1^2 + 2r_1r_2\cos2\delta_1 + r_2^2}{1 + 2r_1r_2\cos2\delta_1 + r_1^2r_2^2} \qquad (7)$$

$$T = \frac{n_1}{n_0} \frac{t_1^2 t_2^2}{\left(1 + 2r_1r_2\cos2\delta_1 + r_1^2r_2^2\right)} \qquad (8)$$

In terms of refractive indices (n_0, n and n_1) the reflectance and transmittance are given by

$$R = \frac{\left(n_0^2 + n^2\right)\left(n^2 + n_1^2\right) - 4n_0 n^2 n_1 + \left(n_0^2 - n^2\right)\left(n^2 - n_1^2\right)\cos2\delta_1}{\left(n_0^2 + n^2\right)\left(n^2 + n_1^2\right) + 4n_0 n^2 n_1 + \left(n_0^2 - n^2\right)\left(n^2 - n_1^2\right)\cos2\delta_1} \qquad (9)$$

$$T = \frac{8n_0 n^2 n_1}{\left(n_0^2 + n^2\right)\left(n^2 + n_1^2\right) + 4n_0 n^2 n_1 + \left(n_0^2 - n^2\right)\left(n^2 - n_1^2\right)\cos2\delta_1} \qquad (10)$$

For absorbing thin film n must be replaced by the complex refractive index $\eta = (n+ik)$. In case of weak absorption with $k^2 \ll (n-n_0)^2$ and $k^2 \ll (n-n_1)^2$, transmittance can be written as [45]

$$T = \frac{16 n_0 n_1 n^2 K}{C_1^2 + C_2^2 K^2 + 2C_1 C_2 K \cos(4\pi n d / \lambda)} \qquad (11)$$

where $C_1 = (n+n_0)(n_1+n)$, $C_2 = (n-n_0)(n_1-n)$ and

$$K = \exp(-4\pi k d/\lambda) = \exp(-\alpha d) \qquad (12)$$

α is the absorption coefficient of the thin film. Generally, outside the region of fundamental absorption (hν>E_g; thin film gap) or of the free carrier absorption (for higher wavelengths), the dispersion of n and k is not very large. The maxima and minima of T in equation (11) occur for

$$4\pi nd/\lambda = m\pi \qquad (13)$$

where m is the order number. In usual case ($n > n_1$, i.e. for a semiconducting film on a transparent non-absorbing substrate, $C_2 < 0$), the extreme values of the transmission are given by the following formulae.

$$T_{max} = \frac{16 n_0 n_1 n^2 K}{(C_1 + C_2 K)^2} \qquad (14)$$

$$T_{min} = \frac{16 n_0 n_1 n^2 K}{(C_1 - C_2 K)^2} \qquad (15)$$

Considering T_{max} and T_{min} as continuous functions of λ, through $n(\lambda)$ and $K(\lambda)$, the ratio of the equations (14) and (15) is given by

$$K = \frac{C_1\left[1-(T_{max}/T_{min})^{1/2}\right]}{C_2\left[1+(T_{max}/T_{min})^{1/2}\right]} \quad (16)$$

The functions $T_{max}(\lambda)$ and $T_{min}(\lambda)$ are shown in the Figure 4.2 as the envelopes of the transmission spectrum. Then from equation (14),

$$n = \left[N+\left(N^2-n_0^2 n_1^2\right)^{1/2}\right]^{1/2} \quad (17)$$

where

$$N = \frac{n_0^2+n_1^2}{2} + 2n_0 n_1 \frac{T_{max}-T_{min}}{T_{max} T_{min}} \quad (18)$$

It is evident from equation (16) that n can be determined from T_{max}, T_{min}, n_1 and n_0 at the same wavelength.

Knowing n, we can determine K from equation (15). The thickness d of the thin film layer can be determined from two maxima or minima using equation (13), by the following formula.

$$d = \frac{M\lambda_1\lambda_2}{2(n(\lambda_1)\lambda_2 - n(\lambda_2)\lambda_1)} \quad (19)$$

where M is the number of oscillations between the two extrema ($M = 1$ between two consecutive maxima or minima); $\lambda 1$, $n(\lambda 1)$ and $\lambda 2$, $n(\lambda 2)$ are the corresponding wavelengths and refractive indices. Knowing d and K we are able to calculate the extinction coefficient k from equation (12).

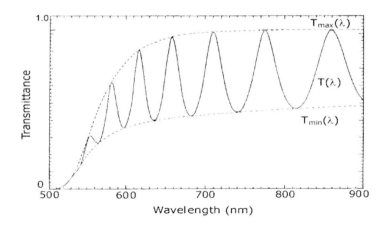

Figure 4.2. A simulated transmission spectrum.

In the fundamental absorption region the transmission T is given by [46]

$$T = A\exp\left(-\frac{4\pi kd}{\lambda}\right) \qquad (20)$$

where

$$A = \frac{16n_0 n_1 (n^2 + k^2)}{\{(n_0 + n)^2 + k^2\}\{(n_1 + n)^2 + k^2\}} \qquad (21)$$

for $k^2 \ll n^2$; the principal variation of T occurs in the exponential term, with $A \approx 1$. Then the expression of T in the absorption edge can be written as

$$T \approx \exp(-\alpha d) \qquad (22)$$

with $\alpha = 4\pi d / \lambda$.

In a semiconductor the electrons can exist in a nearly filled valence band or a nearly filled valence band or a nearly free conduction band. When a photon is incident upon a semiconducting material, its oscillating electromagnetic field may interact with electrons. If the energy of the incident photon is greater than or equal to the energy difference between the conduction band and valence band, then an electron will be transferred from the valence band to the conduction band, provided that no conservation laws are violated by the transition. The probability of transition will determine the probability of a photon to be absorbed. The fundamental absorption refers to the band-to-band transition, i.e. to the excitation of an electron from the valence band to the conduction band. The absorption coefficient $\alpha(h\nu)$ for a given photon energy $h\nu$ is proportional to the probability P_{if} for the transition from the initial state to the final state and to the density of electrons in the initial state, n_i, and also to the density of available (empty) final states, n_f, and this process must be summed for all possible transitions between states separated by an energy difference equal to $h\nu$. $\alpha(h\nu)$ can be expressed as [47]

$$\alpha(h\nu) = A \sum P_{if} n_i n_f \qquad (23)$$

where A is a constant. For absorption transitions between two direct valleys, where all the momentum conserving transitions are allowed, the absorption coefficient is given by

$$\alpha = const.(h\nu - E_g)^{1/2} / h\nu \qquad (24)$$

where E_g is the band gap of the material. In some materials, quantum selection rules forbid direct transitions at momentum $(k)=0$, but allow them at $k \neq 0$. Then absorption coefficient for forbidden direct transitions is given by

$$\alpha = const.(h\nu - E_g)^{3/2} / h\nu \tag{25}$$

In a semiconductor where the conduction band minimum and the valence band maximum occur at different *k* values, optical absorption from the latter to the former require the participation of phonons in order to conserve momentum, because of the change in electron wave vector. Phonons are either emitted or absorbed in this process. The absorption coefficient for an allowed indirect transition is given by

$$\alpha = const.\left[\frac{(h\nu + E_p - E_g)^2}{\exp(E_p/kT) - 1} + \frac{(h\nu - E_p - E_g)^2}{1 - \exp(-E_p/kT)}\right] \tag{26}$$

where the first term represents the contribution of phonon absorption and the second term represents the contribution of phonon emission. E_p is the phonon energy.

Using the transmittance data and Manifacier model [48], absorption coefficients were calculated in the region of strong absorption. A theoretical basis for the UV-visible optical properties of the DLC thin films can be given in terms of the density of states (DOSs) of the material. Useful parameters for characterizing the optical properties of an amorphous material are the optical gap, E_g and the density of states constant B defined [49] by the equation

$$\hbar\omega\sqrt{\varepsilon_2} = A(\hbar\omega - E_g) \tag{27}$$

where A is defined by $A = \log(I/I_o)$ where I_o and I are the intensity of the incident and transmitted beams, respectively and E_g, the optical gap of the material which represents the energy difference between the valence and conduction band and $\hbar\omega$ is the incident photon energy. This equation is often used in the description of the optical properties of amorphous materials and is based on the assumption a parabolic dependence of the DOS on energy [49]. The refractive index and permittivity of a material are the complex numbers $\bar{n} = n + ik$ (k is the extinction coefficient) and are related via $\varepsilon = \varepsilon_1 = i\varepsilon_2$ and are related via $\varepsilon = \bar{n}^2$ so that $\varepsilon = 2nk$. Using the relation

$$\alpha = \frac{4\pi k}{\lambda} = \frac{2\omega k}{c} \tag{28}$$

where α is the absorption coefficients and c, the speed of light, we can convert equ. (27) into

$$(\hbar\omega\alpha)^{1/2} = A(\hbar\omega - E_g) \tag{29}$$

The optical band gap was obtained from Tauc relation as given by equ. (29) can be written as;

$$(\alpha h\nu)^{1/2} = A(h\nu - E_g) \tag{30}$$

Extrapolating the linear portion of the $(\alpha h\nu)^{1/2}$ vs. $h\nu$ plot to the hv axis at $\alpha = 0$, the optical gap has been calculated from the intercept.

5. SYNTHESIS AND OPTICAL PROPERTIES OF DLC THIN FILMS

5.1. Synthesis of Nonmetal and Metal Doped DLC Thin Films

The deposition apparatus for DLC films is a conventional vacuum system equipped with rotary and diffusion pumps. Dow corning silicone oil (DC-704) is used as the diffusion pump fluid. Figure 5.1 shows the schematic diagram of the DC plasma-enhanced chemical vapor deposition (PECVD) apparatus [50]. The vacuum chamber is designed with appropriate stainless steel (SS) vacuum couplings through which different feed-throughs like vacuum port, pressure gauge, gas mixture inlets, thermocouple etc. could be introduced. For the measurement of pressure of the vacuum chamber, pirani and penning gauges are used. The plasma is produced between two parallel plate SS electrodes. The lower disc is grounded upon which the substrate is placed. The upper disc is used as the cathode electrode. The distance between two electrodes can be varied from 0.5 to 3.0 cm. A shutter is used just above the substrate and can be mechanically operated from outside. For substrate heating purpose low tension heating terminals have been connected to substrate heater. Previously calibrated cupper-constantan and chromel-alumel thermocouple were used for measuring the substrate temperature during film deposition. To measure the thickness of the deposited films and its deposition rate, a quartz crystal thickness monitor (HindHivac, Bangalore, Model DTM 101) is attached to the vacuum chamber through the base plate in the conventional manner. A DC power supply, which can be varied from 0 to 5 kV, is used to apply required voltage to the cathode. A photograph of the CVD unit is shown in the Figure 5.2.

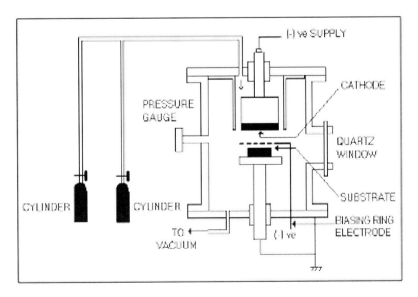

Figure 5.1. Schematic diagram of the DC-PECVD unit.

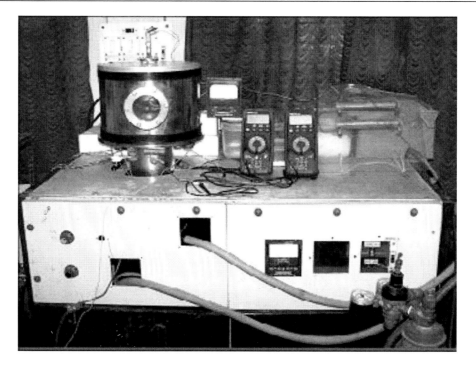

Figure 5.2. The photograph of DC-PECVD unit.

Silicon incorporated DLC (Si-DLC) and fluorine incorporated DLC (F-DLC) films were deposited via DC-PECVD technique. When the chamber pressure attained ~10^{-6} mbar, then acetylene (C_2H_2) gas was introduced and DLC thin films were deposited at a pressure of 0.4 mbar. The substrates (Glass, silicon and alumina) were cleaned by standard cleaning procedure before they were placed in the deposition chamber. For silicon (Si) and fluorine (F) incorporation tetraethyl orthosilicate (TEOS) and hydrofluoric acid (HF) dissolved in methanol solution was used respectively. Ar gas was passed through the solution for bubble formation and then introduced into the chamber with appropriate needle valve arrangement. Si and F at. % was varied in the deposited films by varying the concentration of TEOS and HF in the methanol solution respectively. The deposition conditions for the synthesis of Si-DLC and F-DLC thin films have been shown in Table 5.1.

Table 5.1. Deposition parameters used for Si-DLC and F-DLC thin films synthesis

Deposition parameters	Corresponding values	
	Si-DLC	F-DLC
1. Deposition time	30 min.	20 min.
2. dc voltage	1.0 kV	2.0 kV
3. Electrode distance	2.0 cm	2.0 cm
4. Precursor material	C_2H_2, TEOS	C_2H_2, HF
5. Gas pressure	0.4 mbar	0.15 mbar
6. Substrates used	Glass, Silicon and Alumina	Glass, Silicon
7. Substrate temperature	300 ^{0}C	200 ^{0}C

Table 5.2. Deposition parameters used for Ag-DLC thin films synthesis

Deposition parameters	Corresponding values
1. RF power	150-180 Watt
2. Gases used	CH_4, Ar
3. CH_4/Ar	80-20 %
4. Gas pressure	0.1 mbar
5. Substrates used	Glass, Si (400)
6. Substrate temperature	300 K

The silver incorporated DLC (Ag-DLC) thin films were synthesized by the radio frequency reactive sputtering technique at room temperature (300 K). The silicon substrates were cleaned by standard cleaning procedure before they were placed in the deposition chamber. To deposit the Ag-DLC thin films, the deposition chamber was evacuated to a base pressure 2×10^{-6} mbar and then methane (CH_4) and Argon (Ar) gases were introduced into the chamber for a fixed deposition pressure of 0.1 mbar. A RF power (13.56 MHz) supply was applied in the high purity (99.99 %) Ag sputter target. For the variation of Ag concentration in the films, the ratio of methane to argon sputter gas and RF power has been changed from 80 % to 40 % and from 150 to 180 Watt, corresponding Ag at. % ranging from 0 to 12.5 respectively. The deposition parameters for the synthesis of Ag-DLC thin films have been shown in Table 5.2.

5.2. Optical Properties of Doped DLC Thin Films

Optical transmission spectrum of the film was measured at room temperature by subtracting the transmittance of the glass substrate taking as a reference. It has been observed that optical transmittance increases with the increase of at. % Si in the Si-DLC films as shown in Figure 5.3(a) and the transparency lies in the range of 70 - 80 % in IR and visible range. This increase is certainly related to the increase of sp^3 content in the films due to Si incorporation. Extrapolating the linear portion of the $(\alpha h \nu)^{1/2}$ vs. $h\nu$ plot to the $h\nu$ axis at $\alpha = 0$, the Tauc gap has been calculated from the intercept as shown in Figure 5.3(b). It has been observed from Figure 5.3(c). that the optical gap of the film increased sharply with the incorporation of Si, compared to that of the intrinsic material and lies in the range 0.62 to 2.22 eV for different Si % in the films.

The increase of the optical band gap can be explained by cluster model theory proposed by Robertson et al [51]. The band gap depends on cluster size. In amorphous carbon the sp^3 sites have a wide gap, between σ and σ^* states and sp^2 sites have a variable gap between π-π^* states which depends on the configuration of sp^2 cluster. The increase of silicon and hydrogen content (with increase of TEOS %) in the films sp^3 content increases as confirmed from FTIR and XPS analysis [52], which promote the disordering in the films and lower the sp^2 cluster size. This means Si incorporation enhanced the diamond like behavior and become more transparent with higher optical gap as silicon incorporation increased. Urbach energies calculated from exponential part of the absorption edge provide information about the band edge fluctuations and structural disorder in the films.

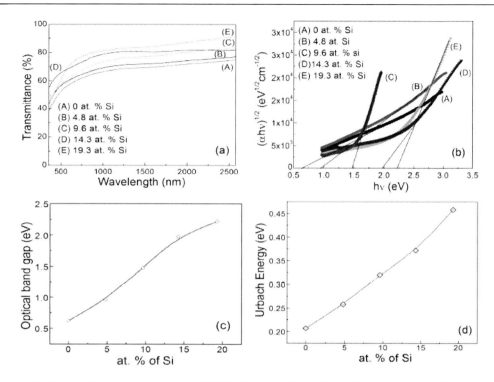

Figure 5.3. Transmittance spectra of the Si-DLC films (a), Tauc plot to determine optical bang gap (b) variation of optical gap with at. % of Si (c) and Variation of Urbach parameter with at. % of Si in the DLC films (d) [Ref. 50].

The Urbach parameter calculated from the optical absorption tails of the films showed that the defect densities increase with Si incorporation in DLC films due to the structural disorder as well as compositional disorder.

Figure 5.4(a) represents the room temperature photoluminescence (PL) emission spectra of Si-DLC films [50]. The PL emission peak of the undoped DLC films was found at 467 nm and the position of the same remains unchanged with silicon incorporation but the intensity increased significantly. This indicates that the room temperature PL is in the visible range. PL peak intensity increases when at. % of Si the DLC films increases as shown in Figure 5.4(b).

In general, features of PL in several amorphous semiconductors such as a-Si:H, a-C:H, and a-C:H:Si are attributed to tail-to-tail states recombination of localized electron-hole pairs. It has recently been suggested that PL arises by the excitation of electrons from π to π^* states in one cluster, which then recombine within the same cluster. Therefore, the PL mechanism of these films can be interpreted using the framework of a:C:H, which is attributed to the radiative recombination of photo-excited electron hole pares in sp^2 bonded clusters. In a-C:H the sp^2 sites form clusters within the sp^3 bonded matrix [53]. The difference between the $\pi-\pi^*$ band gap of the sp^2 sites and the $\sigma-\sigma^*$ gap of the sp^3 sites creates very strong static disorder fluctuations. The π and π^* states lying deeper into the gap form localized tail states. The $\pi-\pi^*$ gap in sp^2 sites are much narrower than the $\sigma-\sigma^*$ gap, and the latter acts as a barrier that strongly localizes the $\pi-\pi^*$ band edge states. Therefore, the electron and holes are closely correlated by coulomb interaction with a short lifetime and the corresponding PL has a strong polarization memory, which can be observed at room temperature [51]. Urbach parameter calculation showed that the disorder increases with increasing the at. % of Si in DLC films.

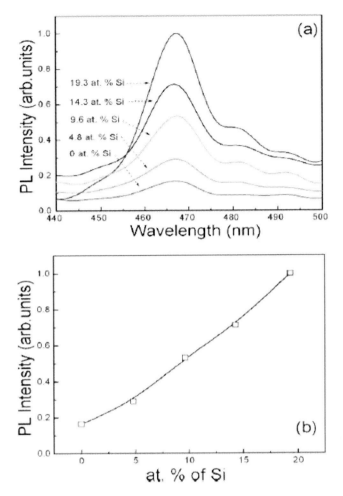

Figure 5.4. PL spectrum of Si-DLC films and (b) change of PL intensity with at. % of Si content in Si-DLC films [Ref. 50].

The broadening of PL as we observed in our films may be attributed to the disorder broadening arising from the broad band tails. It is shown in Figure 5.4(b) that the PL intensities as well as the intensities of C-H$_n$(s) and Si-H$_n$(s) bonds (centered at 2950 cm^{-1} and 1050 cm^{-1} in the FT-IR spectra, respectively) increased with increase the at. % of Si in Si-DLC films. This suggests that the PL of Si-DLC films are dependent on C-H$_n$(s) and Si-H$_n$(s) bonding concentrations (where $n=1,2$). Since the intensity of the PL tends to increase with increasing hydrogen content in the films [54]. Enhancement in photoluminescence properties was attributed to the alteration of the electronic structure by the incorporation of substitutional defect states and the donor activity of silicon.

Figure 5.5(a) shows the transmittance spectra of F-DLC films deposited on glass substrate [55]. It was observed that optical transmittance reduced in the visible and near infrared region with the increase of fluorine concentration in the films. The optical gap of the film decreases from 2.60 to 1.95 eV for the variation of at. % of F in DLC films as shown in. This decrease in the optical band gap would be explained by the fact that the increasing fluorine concentration induces a more significant structural change to highly dominant sp^2

structure. As increase the fluorine content in the films increased, sp^2 content increased but optical gap decreased.

The Urbach energy is one of the standard measures of inhomogeneous disorder in amorphous carbon films. Several factors are responsible for Urbach band tail in semiconductor such as carrier impurity interaction, carrier-phonon interaction, structural disorder etc. Basically this parameter includes the effects of all possible defects. Transitions between extended two localized states in amorphous semiconductors are characterized by the Urbach energy, E_u, which is determined by fitting an exponential function to the slope of the absorption edge. In the energy range $h\nu < E_g$ the optical absorption coefficient normally shows a tail (Urbach band tail) which can be expressed by [56,57]; $\ln\alpha = \ln\alpha_0 - h\nu/E_u$, where α_0 is a constant and E_u is the Urbach energy. The logarithm of the absorption coefficient (α) was plotted as a function of the photon energy ($h\nu$) for F-DLC films deposited with different at. % of F as shown in Figure 5.5(c). The values of the Urbach energy (E_u) for different doping level were calculated by taking the reciprocal of the slopes of the linear portion of these curves. Several factors are responsible for Urbach band tail in semiconductor such as carrier impurity interaction, carrier-phonon interaction, structural disorder etc. Basically this parameter includes the effects of all possible defects. Figure 5.5(d) indicates that the increase of Urbach parameter i.e., defect density, with at. % of F in the DLC films.

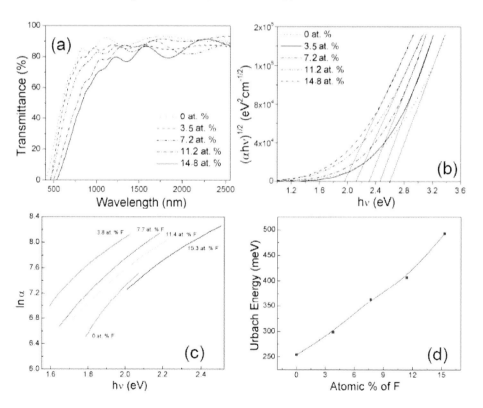

Figure 5.5. Transmittance spectra of the F-DLC films (a), Tauc plot to determine optical bang gap (b), Plot to determine Urbach parameter (c) and variation of Urbach parameter with at. % of F (d) [Ref. 55].

Figure 5.6(a) shows the transmittance spectra of Ag-DLC films deposited on glass substrate [58]. It was observed that optical transmittance reduced in the visible and near infrared region with the increase of Ag concentration in the films. In inset of figure Figure 5.6(a) showed that the optical gap of the film decreases from 2.55 to 1.95 eV for the variation of Ag percentage in DLC films. The optical properties are likely to be controlled by the photonic effects arising from the high density of Ag particles. High resolution transmission electron microscope analysis showed that for low concentration, incorporated Ag atoms were dispersed in the DLC matrix but for high concentration amorphous Ag atoms with diameter ~ 2 nm formed and uniformly distributed with in the DLC matrix [59]. This decrease would be thus associated with the increase in the absorption by the Ag particles in carbon matrix. In addition, the decrease in the optical band gap would be also explained by the fact that the increasing Ag concentration induces a more significant structural change to highly dominant sp^2 structure. The optical band gap decreases with increasing the sp^2 clusters as shown in Figure 5.6(b). According to the cluster model of Robertson the optical band gap of DLC is dependent on the sp^2 cluster size. So, the sp^2 bonds as well as the sp^2 cluster size increases with silver concentration. The DLC films deposited for different at. % of Ag would consist of sp^3 matrix with varying sp^2 cluster concentration as confirmed from XPS analysis [59]. As increases the Ag content in the films, sp^2 content increases while optical gap decreases. Urbach energy, E_u values for different doping level were estimated from the slopes of the linear plot of $\ln\alpha$ against $h\nu$ as shown in Figure 5.6(c) indicating that increase of Ag doping in the DLC films, the urbach parameter i.e., defect density increases.

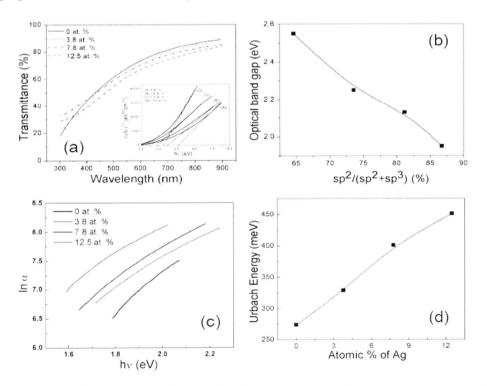

Figure 5.6. Transmittance spectra of the Ag-DLC films and an inset of Tauc plot to determine optical bang gap (a), variation of optical gap with $sp^2/(sp^2+sp^3)$ ratio (b), plot to determine urbach parameter (c) and variation of urbach parameter with at. % Ag in the DLC films (d) [Ref. 58].

CONCLUSION

DLC is a metastable phase of a-C or a-C:H containing a significant fraction of sp^3 hybridization. In spite of having tremendous potential for many applications DLC films possessed many difficulties like having large intrinsic stress and poor electrical conductivity etc. In this article recent progress of DLC and metal and nonmetal DLC films are reviewed due to its tremendous importance in many applications. Nanometer-sized clusters incorporated into the DLC matrix can increase drastically the electrical conductivity, change the optical band gap energy and chemical properties and increase the adhesive property of the film on the material and also resist peeling off the material from the substrate. Recent progress in nonmetal and metal doped DLC nanocomposite films have shown that many of such limitations can be eliminated and new properties can be grafted in the DLC films.

REFERENCES

[1] J. Robertson, *Mater. Sci. Eng. R*, 37 (2002) 129.
[2] W. Jacob and W. Moller, *Appl. Phys. Lett.*, 63 (1993) 1771.
[3] D. R. Mckenzie, *Rep. Prog. Phys.*, 59 (1996) 1611.
[4] P. Koidl, C. Wagner, B. Dischler, J. Wagner and M. Ramsteiner, *Mater. Sci. Forum*, 52 (1990) 41.
[5] M. Weiler, S. Sattel, K. Jung, H. Ehrhardt, V. S. Veerasamy and J. Robertson, *Appl. Phys. Lett.,* 64 (1994) 2797.
[6] J. E. Field, *Properties of Diamond*, Academic Press, London, 1993.
[7] B. T. Kelly, Physics of Graphite, *Appl. Sci. Pub.,* London, 1981.
[8] M. S. Dresselhaus, G. Dresselhaus and P. C. Eklund, *Science of Fullerenes and Carbon Nanotubes,* Academic Press, London, 1996.
[9] J. Robertson, *Adv. Phys.*, 35 (1986) 317.
[10] G. M. Pharr, D. L. Callahan, S. D. Mcadams, T. Y. Tsui, S. Anders, A. Anders, I. G. Brown, C. S. Bhatia, S. R. P. Silva and J. Robertson, *Appl. Phys. Lett.*, 68 (1996) 779.
[11] P. J. Fallon, V. S. Veerasamy, C. A. Davis, J. Robertson, G. A. J. Amaratunga, W. I. Milne, and J. Koskinen, *Phys. Rev. B*, 48 (1993) 4777.
[12] P. C. K. Wagner, B. Dischler, J. Wagner and M. Ramsteiner, *Mater. Sci. Forum*, 52 (1990) 41.
[13] M. Weiler, S. Sattel, K. Jung, H. Ehrhardt, V. S. Veerasamy and J. Robertson, *Appl. Phys. Lett.,* 64 (1994) 2797.
[14] M. F. Ashby and D. R. H. Jonesm, *Eng. Mater. Pergamon* Press, Oxford, 1980.
[15] S. Aisenberg and R. Chabot, *J. Appl. Phys.*, 42 (1971) 2953.
[16] M. A. Tamor and C. H. Wu, *J. Appl. Phys.* 67 (1990) 1007.
[17] P. W. Anderson, *Phys. Rev.* 109 (1958) 1492.
[18] A. Ilie, N. M. J. Conway, B. Kleinsorge, J. Robertson and W. I. *J. Milne, Appl. Phys.,* 10 (1998) 5575.
[19] H. Tsai and D. B. Bogy, *J. Vac. Sci. Technol.* A, 5 (1987) 3287.
[20] S. R. P. Silva and J. D. Carey, *Amorphous Carbon Thin Films*, Handbook of Thin Films, In: H. S. Nalwa (Eds.), Academic Press, New York, 2002.

[21] L. Y. Chen and F. C. Hong, *Appl. Phys. Lett.*, 82 (2003) 3526.
[22] A. Lettington, P. Koidl and P. Oelhafen, Amorphous Hydrogenated Carbon Films, *Proc. Eur. Mater. Res. Soc. Symp.*, Paris, 1987.
[23] D. S. Kim, T. E. Fischer and B. Gallois, *Surf. Coat. Technol.*, 49 (1991) 531.
[24] S. Mitura, E. Mitura and A. Mitura, *Diam. Relat. Mater.*, 302-303 (1995) 4.
[25] V. Elinson, V. V. Sleptsov, A. N. Laymin and A. D. Moussina, *Diam. Relat. Mater.*, 8 (1999) 2103.
[26] T. Hasebe, Y. Matsuoka, H. Kodama, T. Saito, S. Yohena, A. Kamijo, N. Shiraga, M. Higuchi, S. Kuribayashi, K. Takahashi and T. Suzuki, *Diam. Relat. Mater.*, 15 (2006) 129.
[27] Z. Q. Ma and B. X. Liu, Solar Energy Mater. *Solar Cells*, 69 (2001) 339.
[28] A. C. Callegari, K. Babich, S. Purushothaman, S. Mansfield, R. Ferguson, A. Wong, W. Adair, D. Ogrady and V. Chao, *Microelectron. Eng.*, 107 (1998) 4142.
[29] S. R. P. Silva, J. D. Carey, X. Guo, W. M. Tsang and C. H. P. Poa, *Thin Solid Films*, 482 (2005) 79.
[30] S. Bhattacharya, K. Walzer, M. Hietschold, F. J. Richter, *Appl. Phys.*, 89 (2001) 1184.
[31] H. Spicka, M. Griesser, H. Hutter, M. Grasserbauer, S. Bohr, R. Haubner and B. Lux, *Diam. Relat. Mater.*, 5 (1996) 383.
[32] C. L. Tsai, C.F. Chen and C. L. Lin, *J. Appl. Phys.*, 90 (2001) 4847.
[33] M. A. Alaluf and N. Croitoru, *Diam. Relat. Mater.*, 6 (1997) 555.
[34] N. Biswas, H. R. Harris, X. Wang, G. Celebi, H. Temkin and S. Gangopadhyay, *J. Appl. Phys.*, 89 (2001) 4417.
[35] S. Kundoo, P. Saha and K. K. Chattopadhyay, *Mater. Lett.*, 58 (2004) 3920.
[36] C. W. Chen and J. Robertson, *Mater. Res. Soc. Symp. Proc.*, 498 (1998) 31.
[37] J. P. Sullivan, T. A. Friedmann, R. G. Dunn, P. A. Schultz, M. P. Siegal and N. Missert, *Mater. Res. Soc. Symp. Proc.*, 498 (1998) 97.
[38] J. Robertson, *Prog. Solid State Chem.*, 21 (1991) 317.
[39] A. Czyzniewski, *Thin Solid Films*, 433 (2003) 180.
[40] Rusli, S. F. Yoon, H. Yang, J. Ahn, Q. F. Huang, Q. Zhang, Y. P. Guo, C. Y. Yang, E. J. Teo, A. T. S. Wee and A. C. H. Huan, *Thin Solid Films*, 355-356 (1999) 174.
[41] A. Schuler, C. Ellenberger, P. Oelhafen, C. Haug and R. Brenn, *J. Appl. Phys.*, 87 (2000) 4285.
[42] Q. F. Huang, S. F. Yoon, Rusli, Q. Zhang and J. Ahn, *Thin Solid Films*, 409 (2002) 211.
[43] S. Kundoo, P. Saha and K. K. Chattopadhyay, *J. Vac. Sci. Technol. B*, 22(6) (2004) 2709.
[44] O. S. Heavens, *Optical Properties of Thin Solid Films*, Dover Publications, 1965.
[45] J. C. Manifacier, J. Gasiot and J. P. Fillard, J. Phys. E: *Scientific Instruments* 9 (1976) 1002.
[46] J. C. Manifacier, M. D. Murcia, J. P. Fillard and E. Vicario, *Thin Solid Films* 41 (1977) 127.
[47] J. I. Pankove, *Optical Processes in Semiconductors*, Prentice-Hall Inc. New Jersey, 1971.
[48] J. C. Manifacier, M. D. Murcia, J. P. Fillard and E. Vicario, *Thin Solid Films* 41 (1997) 127.
[49] J. Tauc, *Amorphous and Liquid Semiconductors*, New York Plenum, (1974) p. 159.
[50] S. F. Ahmed, *Ph.D. Thesis (2008),* Jadavpur University, Kolkata, India,

[51] Rusli, J. Robertson, G. A. J. Amaratunga, *J. Appl. Phys.* 80 (1996) 2998.
[52] S. F. Ahmed, M. K. Mitra and K. K. Chattopadhyay, *Appl. Surf. Sci.,* 253 (2007) 5480.
[53] J. Robertson, *Diam. Relat. Matter.* 4 (1995) 297.
[54] B. Marchon, J. Gui, K. Grannen, G. C. Rauch, *IEEE Transactions on Magnetics* 33 (1997) 3148.
[55] S. F. Ahmed, D. Banerjee and K. K. Chattopadhyay, *Vacuum*, 84 (2010) 837.
[56] U. N. Maiti, P. K. Ghosh, S. F. Ahmed, M. K. Mitra and K. K. Chattopadhyay, *J. Sol-Gel Sci. Technol.* 41 (2007) 87.
[57] A. E. Rakhshani, *J. Phys. Condens. Matter.* 12 (2000) 4391.
[58] S. F. Ahmed, M. W. Moon and K. R. Lee, *Thin Solid Films*, 517 (2009) 4035.
[59] S. F. Ahmed, M. W. Moon and K. R. Lee, *Appl. Phys. Lett.*, 92 (2008) 193502.

In: Diamonds: Properties, Synthesis and Applications
Editors: T. Eisenberg and E. Schreiner, pp. 55-72
ISBN: 978-1-61470-591-8
© 2012 Nova Science Publishers, Inc.

Chapter 3

POLISHING OF DIAMOND SURFACES

Y. Chen and L. C. Zhang[†]*
[1]School of Mechanical and Manufacturing Engineering,
The University of New South Wales, NSW 2052, Australia

ABSTRACT

Since the successful synthesis of diamond, especially the rapid development of chemical vapour deposition (CVD) diamond technology, diamond has been used extensively in industry and the effective polishing of diamond surfaces has become increasingly important. Over the years, various techniques using a mechanical, chemical or a thermal method, or their synergistic combinations have been developed, of which the dynamic friction polishing (DFP) method is relatively new and appears to be an attractive alternative to provide the efficiency that the conventional methods cannot achieve. In addition, the DFP is an abrasive free technique, requires simple machinery, and can be implemented in a normal ambient environment.

This chapter reviews the latest development on DFP of diamond surfaces, which includes polishing equipments, estimation of interface temperature, exploration of the material removal mechanism, modelling of material removal rate, establishment of polishing map for nanometric surface finish and characterization of surface integrity. It also presents the applications of the method to single crystalline diamond, polycrystalline diamond composites, and CVD diamond films.

1. INTRODUCTION

Since the successful synthesis of diamond, especially the rapid development of chemical vapour deposition (CVD) diamond technology, diamond has been used extensively in industry and the effective polishing of diamond surfaces has become increasingly important. Over the years, various physical and chemical methods have been developed to polish the

---

[*] E-mail:y.chen@unsw.edu.au
[†] Email: Liangchi.Zhang@unsw.edu.au

diamond surfaces. Comprehensive reviews on these techniques have been available in the literature [1-4]. The diamond polishing techniques using a mechanical, chemical or a thermal approach, or their synergistic combinations can be broadly classified into contact and non-contact methods, including mechanical polishing, chemo-mechanical polishing, thermo-chemical polishing, high energy beam (laser/plasma/ion beam) polishing, electrical discharge machining (EDM) and dynamic friction polishing.

Mechanical abrasive polishing has been practised for many centuries, and has been widely used in industry [5-7]. This technique can produce a surface with a roughness of the order of Ra = 0.02 μm without drastically changing the chemical quality of the diamond surface [3-4,8]. The process is straightforward and scalable, and there is no requirement for heating the substrate or using reactive gases. However, the polishing time can be quite long, up to several days, and the applied pressure on the substrates can cause microcracking on CVD diamond films [9]. The polishing rates are extremely low and depend on the quality of the diamond, its lattice orientation and polishing direction.

The chemo-mechanical method [10-14] uses mechanical polishing with the aid of chemicals to enhance the removal rate and to obtain a better surface finish (~ Ra =20 nm) [15-16]. This method exploits the high temperature oxidation property of diamond [3,11] and the compound effect of mechanical abrading, where oxidant etching plays a main role in the material removal process. The processing time is of the order of several hours [17]. However, the heating of the polishing disk and the requirement of oxidizing agents make the polishing process complicated. In addition, to maintain a continuous polishing process, it is essential to remove the chemically reacted products accumulated on the polishing disk.

The thermo-chemical polishing (or hot-metal-plate polishing) is a process using a hot metal plate, and is based on the thermo-chemical reaction between a diamond surface and a hot metal plate at an elevated temperature (730 – 950 °C). It offers a fine surface finish, and the polishing rate is much higher than that of mechanical polishing (in the order of a few μm/h) [18-20]. However, an efficient polishing can only be achieved by heating the polishing disk to a temperature over 750 °C, which requires an evacuated/reductive atmosphere to prevent the metal from oxidation, especially when using iron at high temperatures.

Another thermo-chemical method, diamond etching, uses the principle of diffusion reactions [21]. Metals or alloys (Fe, Mn or molten rare earth metals/alloys such as Ce or La) foils are placed in contact with the diamond films under load at an elevated temperature (600 - 900 °C) in an argon atmosphere [22-27]. An etching rate of approximately 60 μm/min [53] has been achieved by using rare-earth/transition-metal alloys. The advantages of this method are that it is applicable simultaneously to a large number of diamond films, and that it has the flexibility for shaping diamond into non-flat geometries. However, this method does not provide a fine surface finish whose Ra is often of the order of a few micrometers [26].

The high energy beam techniques include Plasma/Ion beam and laser beam polishing. Asperities on a diamond surface are removed by the high energy beam via sputtering, etching, localized heating, high-temperature graphitization and/or oxidation. These non-contact methods generally do not require the application of a force to samples, or do not need to heat them. Hence polishing of non-planar surfaces and/or small areas can be achieved [3,28-31]. Their major disadvantages are the cost of expensive equipment and for maintaining a controlled environment, generally a vacuum environment.

EDM removes materials by harnessing thermal energy produced by pulsed spark discharges which generates a very small plasma channel having a high energy density and a very high temperature (up to 10,000 °C) that melts and evaporates a small amount of diamond surface material. The EDM polishing can provide a very high material removal rate [32-35], but the finial surface finish is only up to a few microns Ra.

Dynamic friction polishing (DFP) is a relatively new method proposed and studied [36-39]. This technique utilizes the thermo-chemical reaction induced by the frictional heating between a diamond specimen and a rotating catalytic metal disk under certain pressure, and enables highly efficient abrasive-free polishing. It appears as an attractive alternative to supplement the deficiency of the conventional polishing methods, because the DFP can obtain very high polishing rate and effectively use the friction energy. The equipment required is simple, and the process can be implemented in a normal ambient environment and does not require a vacuum chamber and/or special heating.

This chapter reviews the latest development on dynamic friction polishing of diamond surfaces, which will include polishing equipments, estimation of interface temperature, exploration of the material removal mechanism, modelling of material removal rate, establishment of polishing map for nanometric surface finish, and characterization of surface integrity. It also presents the applications of the polishing technology in single crystalline diamond, polycrystalline diamond composites (PCDC), and CVD diamond films.

2. POLISHING EQUIPMENTS

A typical DFP process is schematically illustrated in Figure 1. The polishing was conducted by pressing a rotating diamond specimen at a predetermined pressure onto a catalytic metal disk rotating at a high speed in dry atmosphere. The metal disk used in DFP can vary, provided that it consists of catalytic elements. For example, some used nickel or stainless steel [36-37,40] but some others used titanium allay [41] or an intermetallic compound consisting of one or more elements selected from the group of Al, Cr, Mn, Fe, Co, Ni, Cu and Pt and one or more from the group of Ti, V, Zr, Mo, Ta and W [42-44].

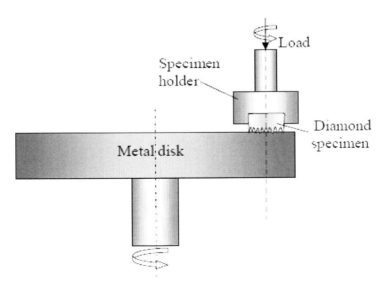

Figure 1. Schematic illustration of a typical dynamic friction polishing.

Another advantage of DFP is that polishing can be carried out on various machines. The polishing metal disk and diamond specimen can be mounted on to a CNC machine centre, universal milling machine, normal polishing or grinding machine, or specifically designed diamond polishing machine for better and stable control of the process. The key requirement is that the equipment needs to provide high combination of pressure and sliding speed to generate adequate dynamic friction heating between diamond and the catalytic metal disk to generate sufficient interface temperature to activate phase transformation or/and chemical reactions during polishing. In addition, the equipment needs to carry out the DFP process efficiently and in a controllable manner to ensure precise and uniform polishing of diamond surfaces.

A wide range of parameters, such as a pressure in the range of 1 to 100 MPa and sliding speed in the range of 10 to 167 m/s at different combinations have been used for polishing. DFP has been conducted at both room temperature and in a heated environment such as 100 to 800 °C. Some techniques were developed by heating the polishing disk or diamond specimens with the intention that the effective polishing can be conducted at a lower pressure or sliding speed.

3. INTERFACE TEMPERATURE

The chemical reactions and phase transformation of diamond play an important role in the material removal of diamond, and these reactions only occur at elevated temperatures. It is therefore important to estimate the temperature during the process. The temperature at polishing interface has been characterized by theoretical modelling and experimental measurement.

Iwai et al used FEM analysis to predict the temperature at single crystalline diamond surface [45]. The results of estimated surface temperature *vs* sliding speed at different pressure are shown in Figure 2. In the simulations, the friction coefficient was selected according to previous experiment, and the high coefficient of 0.2 and low coefficient of 0.08 were used and the results are presented in Figure 2 (a) and (b), respectively.

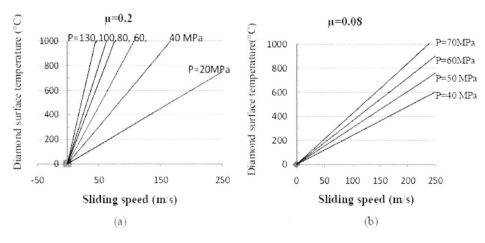

Figure 2. Estimated diamond surface temperature *vs* sliding speed at different nominal pressures [45]. (a) Coefficient of friction μ=0.2. (b) Coefficient of friction μ=0.08.

It can be seen that the higher pressure and sliding speed resulted in higher surface temperature, and the temperature increased lineally with the sliding speed at a given pressure. Since these predicted temperatures are on the nominal diamond surface whose area is much larger than the actual contact area, the predicted temperatures could be lower than the actual interface temperature. The required pressure and speed to achieve the minimum surface temperature of 700 °C for polishing is likely higher than that actually needed.

Chen et al [38] developed a model to predict temperature rise at the interface of the polishing disk and polycrystalline diamond composites (PCDC) asperities. In this model, the Greenwood-Williamson's statistical asperity model was used to characterise the surface roughness of a PCDC specimen. The result was then used to estimate the contact area and total number of contact asperities under an applied polishing load. The heat generated was taken as the product of the frictional force and the relative sliding speed between the asperities and the metal disk surface. Jaeger's moving heat source analysis was then applied to determine the fraction of the heat flowing into the asperities and its counterpart at contact sliding during polishing and to predict the average temperature rise on the contact surface.

Figures 3 shows the variations of the calculated average contact temperature rise with the sliding speed at different nominal pressure applied. The coefficient of friction μ used in the calculation was 0.15. According to these results, the higher values of pressure and sliding speed correspond to a higher heat flux and higher temperature rise. The temperature rise increases with increasing pressure and sliding speed. The dependence of temperature rise on speed appears to be linear for a fixed nominal pressure. Since the model is based on the assumption of no heat loss into the surrounding, the predicted interface temperature rise is the upper bound.

In the current practice, it is almost impossible to measure the interface temperature during diamond polishing. Although the thermocouple technique has been used in temperature measurement [46], fitting a thermocouple into a rotating system of DFP is difficult and the temperature at the polishing interface cannot be measured directly. Thus an attempt was made to measure the PCD subsurface temperature, and then extrapolate it to the polishing surface [47].

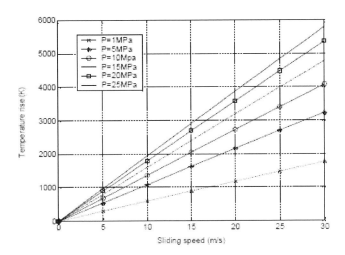

Figure 3. Variation of calculated average temperature rise with sliding speed at different nominal pressures [38].

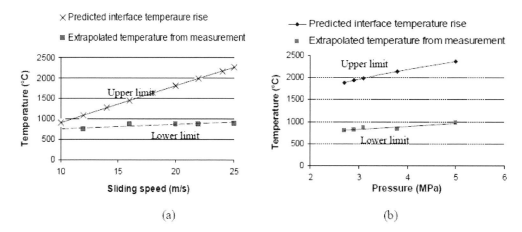

Figure 4. Compare of interface temperature from theoretical prediction and extrapolated from measurement [47]. (a) with speed at pressure 3.1MPa. (b) with pressure at speed 22 m/s.

The temperature picked up by the thermocouple was not at the polishing interface, but at a distance (approximately 0.6 mm) from it. To obtain the interface temperature, a model of a steady temperature in a semi-infinite cylinder [48] was used. The calculation was based on the assumptions that no heat was lost in the process of conduction from the interface to the tip of the thermocouple and no convective heat-losses during the disk/specimen spinning. This is not the same during actual polishing. Hence, an extrapolated interface temperature from the measured result gives the lower bound. The actual interface temperature during polishing is between the lower bound from the experiment and the upper bound from the theoretical prediction.

For comparison, some typical results of the extrapolated interface temperature from the experimental measurement and that from the theoretical prediction are plotted against the variation of sliding speed at a given polishing pressure (Figure 4 (a)) [47], and against the variation of the applied polishing pressure at a given sliding speed (Figure4 (b)). It can be seen that at a given speed, the higher the pressure, the higher the interface temperature. As expected, at any specific combination of sliding speed and pressure, the theoretically predicted temperature rise is always higher than the experimental. Their difference becomes bigger at higher sliding speeds/pressure, possibly due to the stronger convective cooling which was ignored in both the theoretical and experimental modelling.

4. MATERIAL REMOVAL MECHANISMS

The material removal mechanisms of DFP have been studied by a number of research groups. Iwai et al [37] and Suzuki et al [36] investigated the material removal mechanism based on the polishing efficiency in various atmospheres, and carried out x-ray diffraction analyses of the polishing debris and the surface of metal disk. They found that the mechanisms were rapid diffusion of carbon from the diamond to the disk and then evaporation of carbon by oxidization. Huang et al [41] analysed the element composition and chemical state of the diamond film and polishing titanium disk after polishing by using x-ray photoelectron spectroscopy (SPX). They suggested that the chemical reaction between carbon and titanium and the diffusion of carbon atoms into the polishing plate were the material removal mechanisms.

To gain a comprehensive understanding of the polishing mechanism, Chen et al [39-40] combined theoretical and experimental investigations to explore whether chemical reaction and phase transformation had occurred. A theoretical study was carried out with the aid of thermodynamics and chemical kinetics of interface reactions. Scanning electron microscopy (SEM), x-ray diffraction (XRD), Raman spectroscopy, transmission electron microscopy (TEM), electron diffraction, energy dispersive x-ray analysis (EDX) and electron energy loss spectroscopy (EELS) were used to analyse the debris produced during DFP and diamond specimens before and after polishing to clarify the material removal mechanisms.

Figure 5 shows the Raman spectra of diamond specimen surface and polishing debris [39,49]. Non-diamond carbons with Raman peaks at 1585-88 and 1321-22 were detected on the polished surface before cleaning (Figure 5(b), (c)) and in the polishing debris (Figure 5(e)), indicating that transformation did occur during polishing.

Based on the EELS and HRTEM analyses [40], it was found that the polishing debris were mainly of amorphous structure, that included different forms of carbon and iron oxides, etc. From the free energy theory and low-loss energy spectra (Figure 6), the densities of carbon material in polishing debris were calculated to be much less than that of diamond [40]. From high-loss energy spectra, the percentage of sp^2 bonding in the hybridized carbon materials of the polishing debris ranged from 30 to 90%. These results indicate that during DFP, the diamond at surface has transformed to amorphous non-diamond carbon due to the interaction with rotating metal disk at elevated temperature [40].

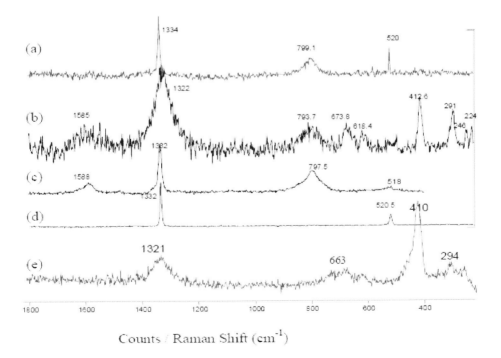

Figure 5. Raman spectra of PCDC specimen surface and polishing debris [39] (a) Before polishing, (b) After polishing with adhered film, (c) After polishing and removal of adhered film, (d) After further polishing with diamond abrasive (e) Polishing debris.

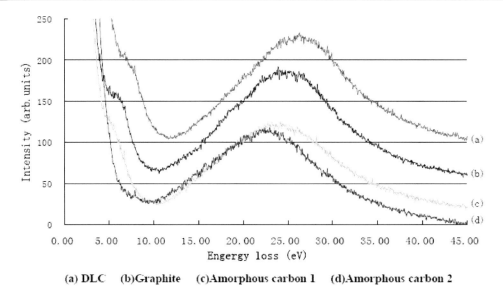

(a) DLC (b) Graphite (c) Amorphous carbon 1 (d) Amorphous carbon 2

Figure 6. Low-energy-loss carbon spectra of polishing debris [40].

After the transformation, the surface of the contact asperities becomes much softer, which can be easily removed mechanically by the continuous sliding between the disk and the diamond surface. SEM and EDX analyses in [39] indicate that carbon was removed with catalyst metals/oxide in particle-like debris.

Additionally, on the contact asperities, diamond and its transformed non-diamond carbon were exposed to oxygen at high temperatures. They could easily react with oxygen and escape as CO and/or CO_2 gas. In addition, the oxidation of carbon would be accelerated by the catalyst metals or metal oxides of Fe, Ni and Cr from the polishing disk. Moreover, diffusion of the carbon from diamond surface to metal disk is another process contributing to the material removal.

Figure 7 summarizes the material removal mechanisms and the associated chemical reactions in the dynamic friction polishing process [40].

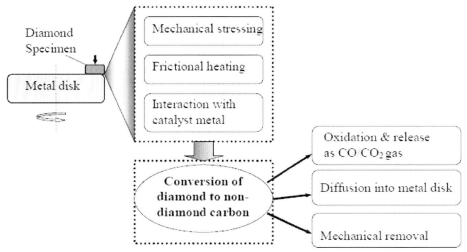

Figure 7. The mechanism map of material removal in dynamic friction polishing [40].

Based on the experimental results and theoretical analyses, the material removal mechanism can be described as follows: conversion of diamond into non-diamond carbon takes place due to the frictional heating and the interaction of diamond with the catalyst metal disk; then a part of the transformed material is removed mechanically from the diamond surface as it is weakly bonded; another part of the non-diamond carbon oxidizes and escapes as CO or/and CO_2 gas; and the rest diffuses into the metal disk. After the phase transformation and the removal of the adhered layer, new asperities would contact with the metal disk, and the transformation and removal are repeated during sliding, and hence result in the continuous removal of diamond.

5. MATERIAL REMOVAL RATE

From the aforementioned understanding, there are various material removal mechanisms involved in DFP and many factors can influence the material removal rate. These include the polishing parameters such as polishing pressure and sliding speed, surrounding gas environment, diamond workpiece material and its initial surface condition, polishing disk material and initial disk temperature etc. By selecting proper polishing conditions, very high material removal rate can be obtained.

Iwai et al [45] studied the material removal rate on DFP of single crystalline diamond, and reported that a very high material removal rate of 520 μm /min (equal to 0.182 mm³/min) was achieved at the polishing speed of 167 m/s and pressure of 100 MPa. The polishing time used was only 0.5 second; this had reduced dramatically from the other diamond polishing method whose polishing time was in the order of hours and days. For CVD diamond, material removal rate of 12 μm/h could be obtained at the polishing speed of 60 m/s and the pressure of 0.31 PMa on a titanium polishing disk [41].

In general, the research found that the pressure and speed needed to be high enough to generate sufficient frictional heating to a critical temperature to trigger the chemical reaction. A high polishing pressure and sliding speed resulted in a high material removal rate. Compared to a normal ambient environment, the material removal rate increases in an oxygen environment but decreases in nitrogen gas [36]. This is because oxidation of carbon accelerates the transformation of diamond to non-diamond carbon and speeds up the material removal in polishing.

Chen et al [50] has systemically studied the material removal rate of two types of thermally stable polycrystalline diamond composites (PCDC). The Type 1 PCDC contains about 75% polycrystalline diamond particles ($C=75\%$) with grain size $\delta \sim 25$ μm (the rest are SiC and Si), and has an initial surface roughness of $\varepsilon = 1.6$ μm. The Type 2 PCDC is of $C = 65\%$, $\delta \sim 6$μm and $\varepsilon = 0.7$μm. The size of both types of PCDC is $D = 12.7$ mm in diameter and 4 mm in thick. To understand the influence of polishing parameters on the material removal rate, the speed V was varied from 8 to 25 m/s for each polishing pressure of 2.2, 2.7, 3.1, 3.8 or 5 MPa (corresponding to load 285, 343, 392, 480 and 637 N) and at a constant polishing time of 3 minutes for Type 1 specimens and 2 minutes for Type 2 specimens, as shown in Figure 8, where the symbols represent the experimental results, the solid lines represent the fitted linear regression lines of the Type 1 specimens, and the dotted one represents those of the Type 2 specimens.

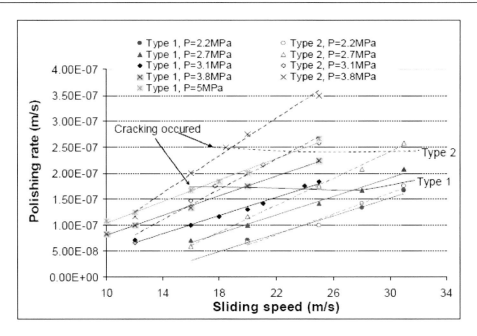

Figure 8. The variation of removal rate with sliding speed at different pressure for both types of PCDC [50].

As shown in Figure 8, at a given sliding speed, the material removal rate increases with the pressure rise. Similarly, a higher speed at a given pressure results in a higher removal rate. For Type 1 specimens, when the sliding speed was lower than 10 m/s, the polishing rate was extremely low and in some cases the material removal was no measurable using the electronic balance available. This means that under these conditions, the temperature rise at interface by sliding friction is not high enough to stimulate the chemical reactions. At a higher sliding speed (>12 m/s), the polishing rate is a function of both the pressure and sliding speed, increasing almost linearly with the speed at every given pressure. For Type 2 PCDC, a similar trend can be seen but with different critical values of the polishing parameters. In this case, when the speed was lower than 12 m/s, the polishing rate was extremely low, sometimes even not measurable.

We can see that in general at an identical sliding speed and pressure, the material removal of the Type 2 PCDC (smaller grain size particles) is higher than that of the Type 1 PCDC. This is because smaller diamond particles have more surface defects and a larger surface area in the composite. The chemical reaction starts at the surface defects, thus reacting faster [51]. However, at a low speed and pressure combination, the Type 1 specimens have a higher material removal rate. This is mainly due to their much greater initial surface roughness (Rmax ≈10 μm) in comparison with the Type 2 PCDC (Rmax ≈ 5 μm) which is a critical factor of temperature rise at the polishing interface [1]. Under such conditions, the material removal is mainly from the surface asperity peaks.

The aforementioned study shows that the material removal of the PCDC in DFP polishing is a function of the constitutive and thermal properties of materials, and also varies with the polishing parameters that generate the frictional heat for chemical reactions. There are multi-variables that influence the material removal rate. Therefore Chen, et al. [52] carried out a dimensional analysis to try to describe quantitatively the material removal thickness d. They found that the following dimensionless formula can fairly accurately estimate d:

$$\frac{d}{D} = 7.39 \times 10^{-26} \left(\frac{\mu L}{ED^2}\right)^{1.42} \left(\frac{VD}{2\chi}\right)^{0.55} \left(\frac{Vt}{D}\right)^{0.64} \left(\frac{C\delta}{D}\right)^{-0.33} \left(\frac{\varepsilon}{D}\right)^{0.43} T_0^{3.95} \tag{16}$$

where the units of the variables are in SI units (kg, m, sec, ^{0}K), and the nomenclatures are shown in Table 1. A comparison of the formula estimation with experimental measurements is presented in Figure 9 [52].

The formula shows that to achieve a higher material removal rate, a greater load L and sliding speed V are generally required; but this is often limited by some technological constraints such as power the consumption allowable and machine capacity available. This difficulty can be overcome by using a smaller L and V but a longer processing time. The formula also indicates that if a PCDC contains larger diamond particles and a higher percentage of diamond, the material removal will be more difficult, while the higher initial disk temperature will increase the material removal rate. The formula can be used as a practical guide for designing a polishing machine or for planning a polishing process to achieve the balance between production rate and performance.

Table 1. Nomenclature

C	composition of diamond in PCDC, %	D	characteristic length of sample, m
d	material removal height, m	E	equivalent Young's modulus, Pa
L	normal load on PCDC specimen, N	q	total heat flux generated by sliding contact, W m^{-2}
T_0	initial temperature, K	t	process time, s
V	sliding velocity, m s^{-1}	δ	characteristic diameter of diamond particle, m
ε	surface roughness	μ	coefficient of friction
χ	thermal diffusivity, m^2 s^{-1}		

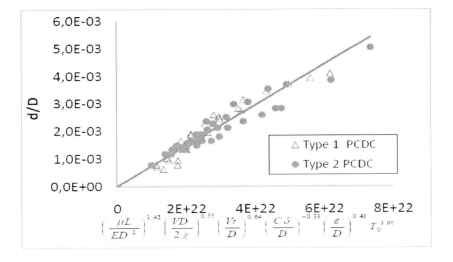

Figure 9. Comparison of model prediction (solid line) and experimental measurements (dots) [52].

6. POLISHING MAP AND SURFACE INTEGRETY

During the dynamic friction polishing of PCDC, although a higher pressure/speed increases the material removal rate, it may also result in cracking [50]. According to Figure8, cracks were observed when the speed-pressure combination is above the solid line for the Type 1 specimens, e.g., under the following polishing conditions: pressure = 5 MPa and sliding speed ≥ 16 m/s; pressure = 3.8 MPa and sliding speed ≥ 20 m/s; pressure = 3.1 MPa and sliding speed ≥ 24 m/s; pressure = 2.7 MPa and sliding speed ≥28 m/s; and pressure = 2.2 MPa and sliding speed ≥31 m/s. For the Type 2 PCDC, cracking occurred under the following polishing conditions above the dotted line as shown in Figure 8: pressure = 3.8 MPa and sliding speed ≥ 18.5 m/s; pressure = 3.1 MPa and sliding speed ≥ 21 m/s, pressure = 2.7 MPa and sliding speed ≥ 31 m/s.

The cracking was likely caused by the non-uniform thermal deformation in the PCDC material in which the coefficient of thermal expansion of diamond (1 × 10^{-6} /K at 300 K [53]) is much lower than that of the binder phase, SiC (3.8 × 10^{-6} /K at 300 K [54]). When temperature increases, the volume expansion of SiC is much larger than that of the PCD. As a result, cracking takes place along the PCD-SiC boundaries when the thermal stresses are large enough, as confirmed by the experimental observations [50].

The above analysis suggests that to avoid cracking, polishing should not be carried out at a very high speed-pressure combination. However, to obtain a reasonable material removal rate, which is a requirement of production, a too low speed-pressure combination is not practical, because the frictional heating at a too low speed-pressure combination cannot generate sufficient temperature rise for chemical reaction and for transforming diamond to non-diamond carbon.

The results in Figure 8 can be more easily visualized as a polishing processing map, as shown in Figure 10 for Type 1 PCDC, a plot of sliding speed *vs* polishing pressure, where the value of the material removal rate (× 10^{-7} m/s) measured at a given pressure and sliding speed is indicated next to the data point [50]. A dotted curve extrapolated through these data show a contour of a constant polishing rate.

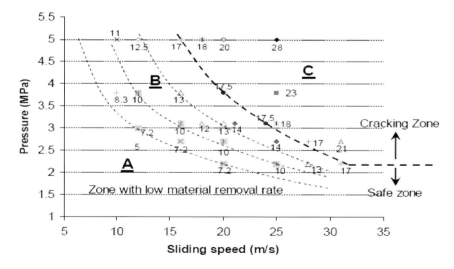

Figure 10. The material removal map [50].

It can be seen from this polishing map that there are three regimes that characterise the dynamic friction polishing of PCDCs. Region *A* is a zone associated with a low or negligible material removal rate and hence is not a practical regime for polishing production. Region *C* is an unsafe zone, in which cracking will occur although the material removal rate can be very high. Region *B* is a safe and workable zone. When a pressure-speed combination falls into this zone, a damage-free polishing with a reasonable material removal rate can be obtained.

For a given pressure (or speed) and a desirable material removal rate, the polishing speed (or pressure) can be easily determined using the polishing map described above. For example, if the desirable polishing rate is 14×10^{-7} m/s, a feasible polishing condition can be taken as speed = 25 m/s with pressure = 2.7 MPa, or speed = 21 m/s with pressure = 3.1 MPa. Using these conditions and further mechanical abrasive polishing was applied to further polish the PCDC, the surface roughness can reach 50 nm Ra in 18 minutes from 1.6 μm Ra (Figure 11), which is more than 10 times faster than the mechanical abrasive polishing process currently used in industry.

In addition, Raman mapping was used to analyse the polished PCDC surface [55]. Figure 12 (a) shows the macro-Raman spectrum which was obtained by averaging the 2943 spectra collected from the rectangular area marked in Figure12 (b). No graphite, metal oxide or non-diamond carbon was detected.

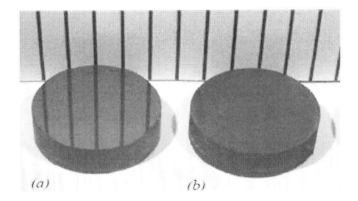

Figure 11. PCD surface [50]. (a) After polishing with mirror finish. (b) before polishing.

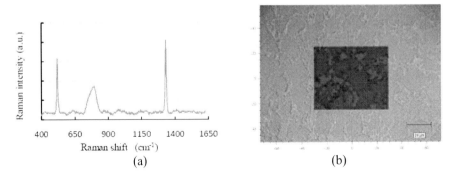

Figure 12. Raman mapping of the final polished PCDC surface [55]. (a) Macro-Raman spectrum (by averaging the 2943 spectra). (b) Raman map of phase distribution (red: diamond, cyan: Si, blue: SiC. No graphite was detected).

The sharp narrow diamond line centred at 1332.4 cm^{-1} indicates a high quality of the completely polished surface. The average spectrum contains deformed amorphous SiC band centre at ~799 cm^{-1} and a high intensity of Si peak at ~521 cm^{-1}. The Raman map (Figure 12(b)) provides the location of the three phases. It can be seen that pristine diamond phase was predominant within the gain areas, as demonstrated in red in Figure12 (b), while SiC and Si (shown as blue and cyan, respectively) distributed along the grain boundaries. All the intermediate soft materials including metal oxide, non-diamond and amorphous graphite which may degrade the quality of the PCDC had been removed from the surface. This verifies that the DFP will not only achieve fast polishing of PCDC, but also maintain high material quality.

7. APPLICATIONS

The DFP method has been used for polishing single crystal diamond, polycrystalline diamond composites (PCDC) and CVD diamond films.

Compared to polishing PCDC and CVD diamond, polishing single crystal diamond using DFP method is the most straightforward. By selecting a proper polishing pressure and sliding speed, e.g., 17 MPa and 35 m/s, a very high polishing rate at 170 μm/min with a high quality surface finish at the roughness ~0.05 μm Ra can be obtained in a minute [56]. This is hundreds times faster than the other polishing techniques reported in the literature [2-3]. In addition, Iwai et al [45] reported that the material removal rate could reach 480 μm/min when polishing at the speed of 167 m/s and pressure of 50 MPa, with an ended surface roughness of Rz = 0.05 μm. This method can be used to manufacture diamond products include diamond jewelry, and to repair worn diamond components such as diamond cutting tools and dressers for grinding wheels.

The polishing became more complicate when applying DFP on thermally stable PCDC which contained diamond, SiC and Si. The two major constituents, diamond and SiC, have very different properties, eg., hardness, coefficient of thermal expansion and chemical reactivity, etc, and the material removal rates for diamond and SiC will be different. Cracking may occur when polishing at a very high speed-pressure combination due to the different coefficient of thermal expansion of diamond and SiC. As detailed in Section 6, by selecting proper polishing parameters, a high polishing rate with a quality surface finish can be achieved in minutes.

DFP has not yet directly been applied to CVD diamond thin films, because the films are often very thin, are of high hardness but weak adhesion to substrate. These make a CVD diamond thin film easy to delaminate, damage and crush during polishing. To date, the DFP has only been applied on the polishing of free stand CVD diamond thick films (~ 0.5 mm) with a low polishing pressure [45]. However, it has shown that DFP is promising in the thick film polishing. For instance, under the condition of speed = 167 m/s and pressure = 1 MPa with blowing oxygen gas, polishing rate obtained was 45 μm/min.

8. SUMMARY

This chapter has reviewed briefly the commonly used techniques for polishing diamond and its related materials. The focus of discussion has been on the latest development on the dynamic friction polishing method, including polishing temperature and its modeling, material removal mechanisms, surface integrity, and material removal rate and its prediction. The discussion has concluded that pressure-speed combinations are the most important. The polishing map developed by Chen, et al [50] provides a useful guide for the design of a DFP process/machine and for a proper selection of polishing parameters. A DFP process is simple, does not require complicated machinery, and can be implemented in a normal ambient environment. By selecting proper polishing parameters, a very high polishing rate with a quality surface finish can be achieved in minutes. The DFP method can be used in polishing single crystalline diamond, polycrystalline diamond and its composites, and CVD diamond thick films. However, its application to diamond thin films is still at infancy.

ACKNOWLEDGMENTS

The authors wish to thank the Australian Research Council for financial support.

REFERENCES

[1] Ralchenko, V.G., Pimenov, S.M., 1998. Processing, in: Perlas, M.A., Popovici, G., Bigelow, L.K. (Eds.), *Handbook of Industrial diamonds and Diamond Films*. Marcel Dekker, New Nork, pp. 983-1021.

[2] Chen, Y., Zhang, L.C., (2009). On the polishing techniques of diamond and diamond composites. *Key Engineering Materials* 407-408: 436-439.

[3] Malshe, A.P., Park, B.S., Brown, W.D., Naseem, H.A., (1999). A review of techniques for polishing and planarizing chemically vapor-deposited (CVD) diamond films and substrates. *Diamond and Related Materials* 8: 1198-1213.

[4] Bhushan, B., Subramaniam, V.V., Gupta, B.K., (1994). Polishing of diamond films. *Diamond Films and Technology* 4: 71-97.

[5] Wilks, J., Wilks, E., 1991. *Properties and applications of diamond*. Butterworth Heinemann.

[6] Field, J.E., 2001. Properties and technology of diamond surfaces, in: Nazare, M.H., Neves, A.J. (Eds.), Properties, growth and applications of diamond. *EMIS Datareviews* Series No. 26 INSPEC IEE.

[7] Hird, J.R., 2002, *The Polishing of Diamond*, Ph.D. thesis, University of Cambridge

[8] Sudarshan, T.S., 1995. *Polishing of diamond films- a review*, in: Jeandin, T.S.S.M. (Ed.), Surface Modification Technologies VIII. The Institute of Materials, pp. 469-481.

[9] Asmussen, J., Reinhard, D.K., 2002. *Diamond Films Handbook*. Marcel Dekker, New York.

[10] Bhushan, B., Subramaniam, V.V., Malshe, A., Gupta, B.K., Ruan, J., (1993). Tribological properties of polished diamond films. *Journal of Applied Physics* 74: 4174-4180.

[11] Ollison, C.D., Brown, W.D., Malshe, A.P., Naseem, H.A., Ang, S.S., (1999). A comparison of mechanical lapping versus chemical-assisted mechanical polishing and planarization of chemical vapor deposited (CVD) diamond. *Diamond and Related Materials* 8: 1083-1090.

[12] Malshe, A.P., Brown, W.D., Naseem, H.A., Schaper, L.W., 1995, *Method of planarizing polycrystalline diamonds, planarized polycrystalline diamonds and products made therefrom,* 5472370.

[13] Hocheng, H., Chen, C.C., (2006). Chemical-assisted mechanical polishing of diamond film on wafer. *Materials science forum* 505-507, pt.2: 1225-1230.

[14] Graebner, J.E., Jin, S., Zhu, W., 1998, *Method and apparatus for chemical-mechanical polishing of diamond*, 5746931

[15] Hsieh, C.-H., Tsai, H.-Y., Lai, H.-T., Lin, H.-Y., 2002. Comparison between mechanical method and chemical-assisted mechanical method for CVD diamond film polishing, in: Malshe, A.P., Maeda, R. (Eds.), *Proceedings of SPIE*.

[16] Cheng, C.Y., Tsai, H.Y., Wu, C.H., Liu, P.Y., Hsieh, C.H., Chang, Y.Y., (2005). An oxidation enhanced mechanical polishing technique for CVD diamond films. *Diamond and Related Materials* 14: 622-625.

[17] Kuhnle, J., Weis, O., (1995). Mechanochemical superpolishing of diamond using NaNO3 or KNO3 as oxidizing agents. *Surface Science* 340: 16-22.

[18] Yoshikawa, M., (1990). Development and performance of a diamond film polishing apparatus with hot metals. Proceedings of SPIE - *The International Society for Optical Engineering* 1325: 210-221.

[19] Tokura, H., Yang, C.-F., Yoshikawa, M., (1992). Study on the polishing of chemically vapour deposited diamond film. *Thin Solid Films* 212: 49-55.

[20] Yoshikawa, M., Okuzumi, F., (1996). Hot-iron-metal polishing machine for CVD diamond films and characteristica of the polished surfaces. *Surface and Coatiins Technology* 88: 197-203.

[21] Jin, S., Graebner, J.E., Kammlott, G.W., Tiefel, T.H., Kosinski, S.G., Chen, L.H., Fastnacht, R.A., (1992). Massive thinning of diamond films by a diffusion process. *Applied Physics Letters* 60: 1948-1950.

[22] Wang, J.Y., Jin, A.Z., Bai, Y.Z., Ji, H., Jin, C.S., (2002). Etching of CVD diamond thick films by rare-earth compound ink. *Journal of Inorganic Materials* 17: 172-174.

[23] McCormack, M., Jin, S., Graebner, J.E., Tiefel, T.H., Kammlott, G.W., (1994). Low temperature thinning of thick chemically vapor-deposited diamond films with a molten Ce---Ni alloy. *Diamond and Related Materials* 3: 254-258.

[24] Jin, S., Graebner, J.E., McCormack, M., Tiefel, T.H., Katz, A., Dautremont-Smith, W.C., (1993). Shaping of diamond films by etching with molten rare-earth metals. *Nature* 362: 822 - 824.

[25] Jin, S., Graebner, J.E., Tiefel, T.H., Kammlott, G.W., Zydzik, G.J., (1992). Polishing of CVD diamond by diffusional reaction with manganese powder. *Diamond and Related Materials* 1: 949-953.

[26] Sun, Y., Wang, S., Tian, S., Wang, Y., (2006). Polishing of diamond thick films by Ce at lower temperatures. *Diamond and Related Materials* 15: 1412-1417.

[27] Johnson, C.E., (1994). Chemical polishing of diamond. *Surface and Coatings Technology* 68-69: 374-377.
[28] Silva, F., Sussmann, R.S., Benedic, F., Gicquel, A., (2003). Reactive ion etching of diamond using microwave assisted plasmas. Diamond and Related Materials, *13th European Conference on Diamond, Diamond-Like Materials, Carbon Nanotubes, Nitrides and Silicon Carbide* 12: 369-373.
[29] Vivensang, C., Ferlazzo-Manin, L., Ravet, M.F., Turban, G., Rousseaux, F., Gicquel, A., (1996). Surface smoothing of diamond membranes by reactive ion etching process. Diamond and Related Materials, *Proceedings of the 6th European Conference on Diamond, Diamond-like and Related Materials Part* 2 5: 840-844.
[30] Buchkremer-Hermanns, H., Long, C., Weiss, H., (1996). ECR plasma polishing of CVD diamond films. *Diamond and Related Materials* 5: 845-849.
[31] Park, J.K., Ayres, V.M., Asmussen, J., Mukherjee, K., (2000). Precision micromachining of CVD diamond films. *Diamond and Related Materials* 9: 1154-1158.
[32] Guo, Z.N., Wang, C.Y., Zhang, F.L., Kuang, T.C., (2001). Polishing of CVD diamond films. *Key Engineering Materials* 202-203: 165-170.
[33] Guo, Z.N., Wang, C.Y., Kuang, T.C., (2002). Investigation into polishing process of CVD diamond films. *Materials and Manufacturing Processes* 17: 45-55.
[34] Guo, Z.N., Huang, Z.G., Wang, C.Y., (2004). Smoothing CVD diamond films by wire EDM with high traveling speed. *Key Engineering Materials* 257-258: 489-494.
[35] Wang, C.Y., Guo, Z.N., Chen, J., (2002). Polishing of CVD Diamond Films by EDM with Rotary Electrode. *Chinese Journal of Mechanical Engineering (in Chinese)* 38: 168-171.
[36] Suzuki, K., Iwai, M., Uematsu, T., Yasunaga, N., (2003). Material removal mechanism in dynamic friction polishing of diamond. *Key Engineering Materials* 238-239: 235-240.
[37] Iwai, M., Uematsu, T., Suzuki, K., Yasunaga, N., (2001). *High efficiency polishing of PCD with rotating metal disc. Proc. Of ISAAT*2001: 231-238.
[38] Chen, Y., Zhang, L.C., Arsecularatne, J.A., Montross, C., (2006). Polishing of polycrystalline diamond by the technique of dynamic friction, part 1: Prediction of the interface temperature rise. *International Journal of Machine Tools and Manufacture* 46: 580-587.
[39] Chen, Y., Zhang, L.C., Arsecularatne, J.A., (2007). Polishing of polycrystalline diamond by the technique of dynamic friction. Part 2: Material removal mechanism. *International Journal of Machine Tools and Manufacture* 47: 1615-1624.
[40] Chen, Y., Zhang, L.C., Arsecularatne, J., (2007). Polishing of Polycrystalline Diamond by the Technique of Dynamic Friction, Part 3: Mechanism Exploration through Debris Analysis. *International Journal of Machine Tools and Manufacture* 47: 2282-2289.
[41] Huang, S.T., Zhou, L., Xu, L.F., Jiao, K.R., (2010). A super-high speed polishing technique for CVD diamond films. *Diamond and Related Materials* 19: 1316-1323.
[42] Jin, Z.J., Yuan, Z.W., Kang, R.K., Dong, B.X., (2009). Study on two kinds of grinding wheels for dynamic friction polishing of CVD diamond film. *Key Engineering Materials* 389-390: 217-222.

[43] Abe, T., Hashimoto, H., Tadeda, S.I., 2003, Grinding and Polishing Tool for Diamond, Method for Polishing Diamond, and Polished Diamond, *Single Crystal Diamond and Single Diamond Compact Obtained Thereby,* 6,592,436.

[44] Neogi, J., Neogi, S., 2008, *Apparatus and method for polishing gemstones and the like.*

[45] Iwai, M., Takashima, Y., Suzuki, K., Uematsu, T., 2004. Investigation of polishing condition in dynamic friction polishing method for diamond, *7th International Symposium on advances in abrasive technology*, Buras, Turkey, pp. ISAAT 2004-2047(2001-2006).

[46] Kennedy, F.E., Frusescu, D., Li, J., (1997). Thin film thermocouple arrays for sliding surface temperature measurement. *Wear* 207: 46-54.

[47] Chen, Y., Zhang, L.C., Arsecularatne, J., (2008). Temperature Characterization for Nano-polishing of PCD Composites. *Key Engineering Materials* 381-382: 513-516.

[48] Carslaw, H.S., Jaeger, J.C., 1959. *Conduction of heat in solids*, 2nd ed. Clarendon Press, Oxford.

[49] Chen, Y., Zhang, L.C., Arsecularatne, J.A., (2007). Polishing of Diamond Composite Cutting Tools. *International Journal of Surface Science and Engineering* 1: 360-373.

[50] Chen, Y., Zhang, L.C., (2009). Polishing of Polycrystalline Diamond by the Technique of Dynamic Friction, Part 4: Establishing the Polishing Map. *International Journal of Machine Tools and Manufacture* 49: 309-314.

[51] Lee, J.-K., Anderson, M.W., Gray, F.A., John, P., Lee, J.-Y., Baik, Y.-J., Eun, K.Y., (2004). Oxidation of CVD diamond powders. *Diamond and Related Materials* 13: 1070-1074.

[52] Chen, Y., Nguyen, T., Zhang, L.C., (2009). Polishing of Polycrystalline Diamond by the Technique of Dynamic Friction, Part 5: Quantitative analysis of material removal. *International Journal of Machine Tools and Manufacture* 49: 515-520.

[53] Nepsha, V.I., 1998. Heat capacity, conductivity, and the thermal coefficient of expansion, in: Prelas, M.A., Popovici, G., Bigelow, L.K. (Eds.), *Handbook of industrial diamonds and diamond films.* Marcel Dekker, New York.

[54] Goldberg, Y., Levinshtein, M.E., Rumyantsev, S.L., 2001. *Properties of Advanced SemiconductorMaterials GaN, AlN, SiC, BN, SiC, SiGe.* John Wiley and Sons, Inc., New York,.

[55] Chen, Y., Zhang, L.C., 2010. Quality Verification of Polished PCD Composites by Examining the Phase transformations, *IUTAM Symposium on Surface Effects in the Mechanics of Nanomaterials and Heterostructures.* Springer, Beijing, China.

[56] Chen, Y., Zhang, L.C., (2010). Fast Polishing of Single Crystal Diamond. Advanced Materials Research 97-101 4096-4099.

In: Diamonds: Properties, Synthesis and Applications
Editors: T. Eisenberg and E. Schreiner, pp. 73-91
ISBN: 978-1-61470-591-8
© 2012 Nova Science Publishers, Inc.

Chapter 4

ELECTROCHEMICAL OXIDATION OF ORGANIC POLLUTANTS IN AQUEOUS SOLUTIONS USING BORON-DOPED DIAMOND ANODES: CYCLIC VOLTAMMETRIC BEHAVIOR

Nasr Bensalah[1,2,*] *Aline S. Sales*[3] *and Carlos A. Martínez-Huitle*[3,†]

[1]Department of Chemistry, Faculty of Sciences of Gabes,
University of Gabes, Cite Erriadh, Zrig 6072, Gabes, Tunisia
[2]Department of Chemical Engineering, Texas AandM University at Qatar,
Education City, PO Box 23874, Doha, Qatar
[3]Universidade Federal do Rio Grande do Norte, Instituto de Química, Campus
Universitário - Lagoa Nova - CEP 59.072-970; Natal/RN - Brazil

ABSTRACT

In recent years boron-doped diamond (BDD) has emerged as a versatile and sensitive electrode material and has been used in an increasing number of applications. This article reviews some of the factors which influence its performance as an electrode, in particular focusing on cyclic voltammetric behavior during electrochemical oxidation process. The typical electrochemical response of a BDD electrode in presence or absence of organic pollutants in solution is discussed and how this relates to complexities in its process performance. The implications for the applications of BDD and prospects for the future are discussed.

1. INTRODUCTION

In recent years boron-doped diamond (BDD) has emerged as a versatile and sensitive electrode material with suggested applications in electroanalysis [1], biological sensing [2, 3],

---
[*] E-mail: nasr.bensalah@issatgb.rnu.tn and/or nasr.bensalah@qatar.tamu.edu
[†] E-mail: carlosmh@quimica.ufrn.br

electrosynthesis [4], wastewater treatment [4, 5, 6] and drinking water disinfection [7]. The first reports of the use of BDD as an electrode date from about 15 years ago and since that time an extensive literature has grown [8], resulting in increased understanding of the electrochemical properties of the material and its potential applications. While at one time this material was only available to the handful of research groups world-wide with facilities to synthesize it by chemical vapor deposition (CVD), BDD is now available from a variety of commercial sources and is more routinely used.

BDD anodic oxidation became a promising technique for wastewaters treatment and it was largely applied for the treatment of organics in laboratory and bench-scale. BDD electrodes exhibit high overpotentials for the oxygen and hydrogen evolution resulting in a wide electrochemical potential window which aids in the analyses of numerous chemical species. In previous results, electrochemistry of BDD and its advantages compared to conventional electrodes have been detailed for electrochemical oxidation of several important chemical species. Compared with other electrode materials, BDD has revealed a higher stability and efficiency. This fact is due to the synthetic BDD thin film high chemical and dimensional stability, low background current and a very wide potential window of water stability.

Depending on the interactions between hydroxyl radicals and the electrode surface, anodes can be distinguished as "active" or "non-active" materials [4, 5, 7]. Among the particular characteristics of BDD, its chemical inertness is well recognized; for this reason, this material is considered as an ideal "non-active" electrode on which organics oxidation and oxygen evolution take place through the formation of weakly adsorbed and very reactive hydroxyl radicals.

Anodic oxidation of organic matter with platinum and metal oxides electrodes often leads to the formation of polymers that disable electrode; this problem does not arise with BDD electrode if it operates at high current densities able to generate strong oxidants as hydroxyl radicals. BDD-anodic polarization leads to the electrogeneration of high amounts of hydroxyl radicals from water discharge at the anode surface. These hydroxyl radials are assumed to be free, i.e., not adsorbed on the electrode surface which helps to increase the efficiency of the electrochemical process by mediating the oxidation in the vicinity of the electrode surface. The generation of these strong oxidants allows considering BDD-anodic oxidation as an advanced oxidation process. The production of these radicals is reported on other electrode materials such as SnO_2 and PbO_2, but the use of these electrodes in the treatment of water contaminated with organic matter has not proven to be highly effective.

Furthermore, it is reported that in addition to the formation of hydroxyl radicals, other inorganic oxidants such as persulphate, perphosphate and hypochlorite can be electrogenerated from electrolyte oxidation on BDD anodes [4, 5, 7, 9]. These oxidants can complement the direct and mediated by hydroxyl radicals' oxidation mechanisms. The oxidation of organic compounds on the electrode BDD can happen in three competitive pathways: oxidation at the surface of the electrode and mediated oxidation close to the electrode surface by hydroxyl radicals generated from water discharge and mediated oxidation in the bulk of solution by other strong oxidants formed at the anode from supporting electrolyte oxidation.

One objective of this article is therefore to summarize the factors which influence the performance of BDD as an electrode, to allow the more 'casual user' to understand the material without having to perform an extensive literature search. This review is intended to provide an overview of factors which influence the electrochemical properties of diamond and in particular the relationships between cyclic voltammetric behavior, electrochemical activity and electrolysis process. There are many interesting aspects of diamond electrochemistry which are outside the scope of this review but are addressed elsewhere. In particular there are already several excellent reviews on the use of BDD in electroanalysis, synthesis of diamond and diamond applications [1-8]. The many successful applications of BDD will therefore not be addressed here. However the interested reader is directed elsewhere for more detailed information on these subjects [9]. Although other forms of diamond exhibit conductivity and electrochemical activity, this review will focus on predominantly on the electrochemical properties of BDD. Therefore, the goal of this work was to enhance understanding on electrochemical oxidation of organics on BDD anodes using cyclic voltammetry and galvanostic electrolyses. Aiming to check the influence of chemical structure on the electrochemical behavior of organic pollutants, different chemical structures were chosen such as phenols, dyes, surfactants and glycols. The degradation and mineralization of these compounds were proved by analytical measurements of chemical oxygen demand (COD), total organic carbon (COT), inorganic carbon (CO_2), chromatographic analyses in some cases and UV-visible spectrophotometry especially in the case of dyes. Two mechanisms will be distinguished for the electrochemical oxidation of organic compounds on BDD: direct oxidation and indirect oxidation via electrogenerated intermediates formed at high anodic potentials. Finally, this is not intended as an exhaustive review of all literature on BDD [7-35], which is immense, but as an introduction to the general principles, cyclic voltammetry and galvanostic electrolyses, which influence the performance of BDD as an electrode material.

2. EXPERIMENTAL SECTION

Chemicals. Ethylene glycol, glycerol and diethylene glycol and the inorganic reagents are purchased from Fluka and used as received. MB and RB were of analytical grade and purchased from Fluka. The other chemicals such as Na_2SO_4, NaCl, Na_3PO_4, $K_2Cr_2O_7$, H_2SO_4, NaOH, $HgSO_4$ and Ag_2SO_4 are of analytical grade and purchased from Fluka or Merck. All solutions were prepared with deionized water having $18 m\Omega^{-1} cm^{-1}$ resistivity from a Mill-Q™ system.

Boron doped diamond electrode: In this case, thin BDD films were synthesized by Adamant Technologies (Neuchatel, Switzerland). It was grown onto a conductive p-Si substrate (0.1 Ω cm, Siltronix) via the hot-filament, chemical-vapor-deposition technique (HF-CVD) [21]. This procedure gave a columnar, randomly textured, polycrystalline diamond coating, with a thickness of about 1 μm and a resistivity of 15 mΩ cm (±30%). The diamond electrode was place on a metal support (Ti) and electrical connection was made scratching the backside of Si substrate and then coating the area with Ag paste. The Ti support was polished clean prior to contact. Subsequently, Ti support and the borders of BDD surface were covered with an isolator polymer to protect the electrode device when it was immersed in the aqueous solutions.

Morphological and electrochemical characterization of BDD film: As a first step, we have characterized typical BDD surface topologies by atomic force microscopy (AFM); a

Nanoscope IIIa Scanning Probe microscope controller connected with a nanoscope multimode SPM, both from Digital Instruments were adopted for AFM analysis and obtained its cyclic voltammogram in H_2SO_4 as supporting electrolyte (scan rate 100 mV s^{-1}). In this case, potential values were referred to a saturated calomel electrode (SCE).

Analytical procedure. The carbon concentration was monitored using a Shimadzu TOC-5050 analyzer. Chemical Oxygen Demand (COD) was determined using a HACH DR200 analyzer. Carboxylic acids were monitored by HPLC using a Supelcogel H column (mobile phase, 0.15% phosphoric acid solution; flow rate, 0.15 ml min^{-1}). In this case the UV detector was set at 210 nm. Aromatics were monitored by HPLC using a Nucleosil C18 column (mobile phase 60 % water - 40 % methanol; flow rate, 0.50 ml min^{-1}). The UV detector was set to 270 nm. Carboxylic acids were monitored by HPLC using a Supelcogel H column (mobile phase, 0.15% phosphoric acid solution; flow rate, 0.15 ml min^{-1}). In this case the UV detector was set at 210 nm. UV-visible spectra were obtained using a Shimadzu 1603 spectrophotometer and quartz cells.

Determination of Average Current Efficiency (ACE). The chemical oxygen method was used for the determination of the current efficiency during the BDD anodic oxidation of organics. In this method, the COD was measured during electrolysis and the instantaneous current efficiency (ICE) was calculated using the relation:

$$ACE = \frac{[COD_0 - COD_t] \, F \, V}{8 \, I \, t}$$

where COD^0, COD_t and $COD_{t+\Delta t}$ are the initial chemical oxygen demand (in gO$_2$ dm^{-3}) and the chemical oxygen demand at times t and t (in seconds), respectively, I is the current intensity (A), F is the Faraday constant (96487 C mol^{-1}), t is the time (in seconds), V is the volume of the electrolyte (dm^3) and 8 is a dimensional factor for unit consistence

$$\left(\frac{32 \text{ g } O_2 \cdot \text{mol}^{-1} \, O_2}{4 \text{ mol } e^- \cdot \text{mol}^{-1} \, O_2} \right).$$

Determination of the mean oxidation state of carbon. The mean oxidation state of carbon (MOSC) can be calculated from the values of TOC and COD using Eq. (1):

$$MOSC = 4\left(1 - \frac{COD}{TOC}\right) \qquad (1)$$

where COD is the chemical oxygen demand (in molO$_2$ L^{-1}) and TOC is the total carbon (in mol C L^{-1}), respectively. This parameter oscillates between −4 for the minor oxidation state (CH$_4$) and +4 for the major oxidation state (CO$_2$).

Voltammetry experiments. Electrochemical measurements were obtained using a conventional three-electrode cell in conjunction with a computer-controlled potentiostat/galvanostat (Autolab Model PGSTAT 30, Eco Chemie B.V., Utrecht, Netherlands). Diamond was used as the working electrode, Hg/Hg$_2$Cl$_2$.KCl (sat) as a reference and platinum as a counter electrode. The BDD electrode was circular (25 mm diameter) with a geometric area of 1 cm^2. Voltammetry experiments were performed in unstirred solutions (100 ml). Anode was anodically polarized during 5 minutes with a 1 M H₃PO₄ solution at 0.1 A prior to each experiment.

Galvanostatic electrolyses. The galvanostatic electrolyses were carried out in a single-compartment electrochemical flow cell. Diamond-based material (Adamant Technologies, Switzerland) was used as anode and stainless steel (AISI 304) as the cathode. Both electrodes were circular (100 mm diameter) with a geometric area of 78 cm² each and an electrode gap of 9 mm. The electrolyte was stored in a glass tank (500 ml) and circulated through the electrolytic cell by means of a centrifugal pump. The electrolyte flow rate through the cell was 2500 cm³min⁻¹. The mass transport coefficient of the electrolytic system was calculated using the system ferro/ferri cyanide and it was equal to 2.10^{-5} m.s⁻¹. A heat exchanger was used to maintain the temperature at the desired set point. The experimental setup also contained a cyclone for gas–liquid separation, and a gas absorber to collect the carbon dioxide contained in the gases evolved from the reactor into sodium hydroxide. During the experiments the pH was monitored but not controlled.

3. RESULTS AND DISCUSSION

3.1. Synthesis and Characterisation of BDD Films

The morphology of BDD films can be observed using Atomic Force Microscopy (AFM) which reveals information about crystal orientation and size. Figure 1 shows an AFM image of a typical commercial microcrystalline film, with well defined crystal facets clearly observable with a range of orientations and size in the order 10 – 50 μm. Crystal faces which appear square in shape are (100) grains, while (111) grains appear triangular. Typically, these films consist of nodular features of about 1 μm in diameters which are in turn made up of much smaller nano-grains of 20 – 100 nm.

3.2. Electrochemical Characterization

Figure 1b shows the cyclic voltammograms for acidic media on diamond electrode.

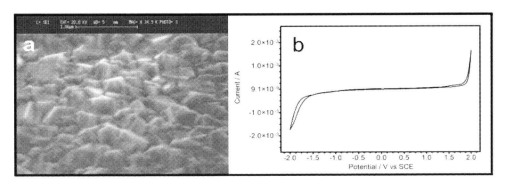

Figure 1. BDD images carried out by the AFM technique and CV curve for BDD electrode, in acidic media (0.5 M H₂SO₄). BDD area of the electrode was 0.78 cm² and scan rate: 100 mV s⁻¹.

The cyclic voltammetric curves for the diamond electrode obtained after purging with nitrogen gas is featureless, and the background current is very low (lees than 1<A). In the acidic media the oxygen evolution peak appear about +2.0V vs SCE (curve A).

3.3. Cyclic Voltammetry

Figure 2 presents oxidative cyclic voltammograms on BDD anodes of aqueous solutions polluted with four organic compounds (cathecol, methylene blue, sodium dodecylbenzene sulfonate and ethylene glycol). As it can be seen the presence of aromatic compounds has led to the appearance of at least one anodic peak, but in presence of aliphatic compounds no anodic peaks were observed. In the later case, a shift of oxygen evolution to more anodic potentials was observed.

These results show that aromatic or aliphatic nature of pollutant influences largely the voltammetric behavior of organics and can be explained in terms of electronic densities which are higher in aromatics than in aliphatic compounds. The highest electronic density, the more easy direct electrons transfer between electrode and organic compound. Furthermore, number, location and irreversibility (absence of cathodic peaks in reverse scan) of anodic peaks evidenced involvement of EC (E: electron transfer and C: chemical reaction) mechanisms occurring in several successive electrochemical and chemical steps.

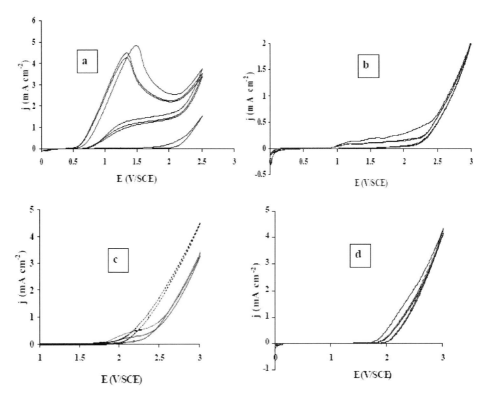

Figure 2. Oxidative cyclic voltammograms on BDD anodes of aqueous solutions polluted with four organic compounds (cathecol, methylene blue, sodium dodecylbenzene sulfonate and ethylene glycol).

On the other hand, table 1 illustrates the values of first peak potentials observed on the voltammograms of several organic molecules in water. It is clear that the nature of the organic compound and the presence of conjugated system in its chemical structure have a great influence on its electrochemical behavior on BDD anodes. More the electron density on molecule is high, easier it is to extract electrons from its highest occupied molecular orbital (HOMO). Dyes which have more extended conjugated systems have the lowest first peak potentials, but glycols which their structures do not present any conjugated systems are the most difficult to be oxidized on BDD anodes.

Table 1. Voltammetric results obtained during the electrochemical oxidation on BDD anodes of phenols, anilines, dyes, surfactants and glycols

	Organic compound	Peak potential (V/SCE)	References
Phenols	Benzene	2.00[a]	[17]
	Phenol	1.45[b]	[18]
	Hydroquinone	1.30[c]	[10]
	Cathecol	1.35[c]	[10]
	Resorcinol	1.55[c]	[10]
	Pyrogallol	0.95[d]	[29]
	1,2,4-trihydroxybenzene	1.00[d]	[13, 14, 15]
	4-nitrocathecol	1.40[d]	This work
	4-chlororesorcinol	1.25[d]	This work
	2,4,6-trinitrophenol	2.20[d]	[10]
Anilines	Aniline	1.10[d]	This work
	4-chloroaniline	0.55[d]	This work
	4-nitroaniline	1.15[d]	This work
	2,4-dinitroaniline	1.90[e]	This work
Aromatic acids	Benzoic acid	2.25[e]	[34]
	Salicylic acid	1.80[e]	[34]
	Phtalic anhydride	Absence of anodic peaks[e] in the first scan but in second scan 2 new peaks are observed	[34]
Dyes	Methylene Blue	1.25[d]	This work
	Methyl Orange	0.95[d]	This work
	Eriochrome Black T	2.10[d]	[35]
	Congo Red	0.95[d]	This work
	Rhodamine B	1.00[d]	This work
	Alizarin Red S	0.85[d]	This work
Surfactants	Sodium dodecylbenzene sulfonate	2.00[d]	[28]
Glycols	Ethylene glycol	Absence of anodic peaks[c,d,e] during more than 5 consecutive scans	[30]
	Glycerol	Absence of anodic peaks[c,d,e] during more than 5 consecutive scans	[30]
	Diethylene glycol	Absence of anodic peaks[c,d,e] during more than 5 consecutive scans	[30]

[a]: H2SO4 (0.5 M); [b]: HClO4 (1 M); [c]: Na3PO4 (0.1 M), pH=2; [d]: Na2SO4 (0.1 M), pH=2; [e]: NaClO4 (0.1 M), pH=2

In the case of aromatics, it is clear that the electronic effects of substituent groups have influence on the direct oxidation of these compounds on BDD anodes. Generally, withdrawing groups presence lead to more difficult oxidation, whereas the presence of donating groups facilitates the oxidation. This can be observed in the cases of: 2,4,6-trinitophenol and phenol; benzoic, salicylic acids and phtalic anhydride; aniline, 2-chloroaniline, 2-nitroanaline and 2,4-dinitroaniline. However, in some cases, it is difficult to observe the electronic effects on the first peak potential especially when more than two groups are present on the benzene ring like in the cases of dyes and surfactants.

In the case of phenols, it is remarkable that number and position of hydroxyl groups on the benzene ring have a great influence on the fist peak potential of phenols. Oxidation of benzene is more difficult than that of all phenols which is related to electronic effects of hydroxyl groups. Moreover, when the number of hydroxyl groups on the benzene ring increases the oxidation seems to be easier. In fact, phenol is also more difficult to be oxidized than polyhydroxybenzenes. In the case of dihydroxybenzenes (Bensalah et al., 2005), it seems that the position of the two hydroxyl groups facilitates further the oxidation in the order: para> ortho> meta.

During voltammetry experiments, the anodic current peaks decrease in the subsequent cycles. This observation can be explained as a result of fouling phenomenon. The activity of the electrode surface becomes less due to this passivating behavior. However, this fouling layer can be removed by anodic polarization in the potential region of water decomposition (>2.3 V). This has already been confirmed by some earlier investigations where the activity of BDD surface was restored by anodic polarization method.

3.4. Cyclic Voltammetric Behavior: Brief Overview

Nasr et al. [10] studied the cyclic voltammetric behavior of solutions containing 13.5 mM of catechol, resorcinol, and hydroquinone and 3333.33 mg dm^{-3} of Na$_3$PO$_4$ at pH 2 using BDD electrodes, as is shown in Figures 5 (potential range from -2.0 to 2.5 V vs SCE) and Figure 6 (potential range 0.0-2.5 V vs SCE). In both figures, the curve obtained under the same experimental conditions in presence and absence of organic matter. As it can be observed, the voltammetric behavior is very similar for catechol and hydroquinone. For both isomers, an anodic oxidation peak with a similar current was observed at about 1.35 V vs SCE. This peak has a reverse peak at about -0.6 V for catechol and about -0.8 V for hydroquinone. The size of the cathodic peaks was smaller than that of the anodic, suggesting electron transfer mechanisms. The anodic oxidation peak decreases in size with the number of scans. This decrease is important in the voltammograms with 0.0V as the lower potential, and smaller in those with a lower potential value of -2.0 V vs SCE. The cathodic peak increases slightly with the number of cycles in the case of hydroquinone and is maintained in the voltammograms of catechol. The hydrogen evolution starts at higher overpotentials with the presence of both catechol and hydroquinone. The oxygen evolution does not seem to be affected by the presence of catechol. For hydroquinone, a significant shift toward the right is observed in the oxygen evolution. On the other hand, the voltammetric behavior of resorcinol is completely different from that shown by the other two isomers. The current densities are smaller, and no reverse peak is observed. Likewise, the anodic peak decreases strongly in both figures. In the voltammograms of Figure 6 (lower potential value 0.0 V vs SCE), it can be observed that the oxidation peak is overlapped with others in the second and in the third scans. The hydrogen evolution is shifted toward higher overpotentials more markedly for this compound, and oxygen evolution is not affected by its presence.

To justify the differences observed between the three isomers, two points were considered: (1) Hydroquinone and catechol have a reversible quinonic form. This justifies the presence of a cathodic reduction peak in their voltammograms. The quinonic compound generated in the first stage of the electro-oxidation of resorcinol is not thermodynamically stable, and thus no reduction peak is observed. (2) The reactivity of the aromatic ring activated with an OH group increases in the ortho- and para-positions. Thus, hydroquinone and catechol have the entire aromatic ring activated, while carbon 5 of resorcinol is not activated. This can justify its lower reactivity.

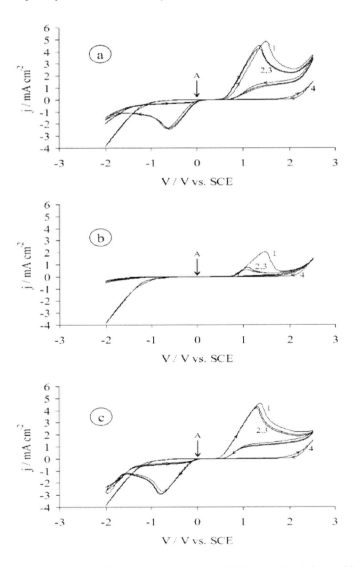

Figure 5. Cyclic voltammograms on BDD anodes of catechol (a), resorcinol (b), and hydroquinone (c) solutions (13.5 mM) on sodium phosphate/phosphoric acid (pH 2, 3333.33 mg dm-3). (1) First cycle; (2) second cycle; (3) third scan; and (4) nonorganic matter. Auxiliary electrode: Platinum. Reference electrode: SCE. Scan rate: 100 mV s-1. A: Anodic start of the CV. Adapted from [10].

The decrease in the peak size in the voltammograms of Figure 6 can be justified in terms of adsorption of the dihydroxybenzene onto the BDD surface and also in terms of polymer formation. In the voltammograms with 0.0 V vs SCE as the lower potential, the enolic form cannot be regenerated after each cycle and some active sites remain occupied by the quinonic form. Thus, the peak current obtained is smaller, and it will decrease in size with each cycle. When the lower potential is -2.0 V vs SCE (Figure 5), some of the active sites become occupied by the enolic form (after the reduction of the corresponding benzoquinone), and thus the peak size is almost maintained.

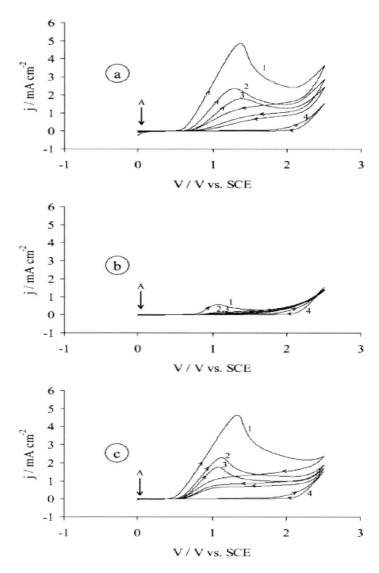

Figure 6. Cyclic voltammograms on BDD anodes of catechol (a), resorcinol (b), and hydroquinone (c) solutions (13.5 mM) on sodium phosphate/phosphoric acid (pH 2, 3333.33 mg dm-3). (1) First cycle; (2) second cycle; (3) third scan; and (4) nonorganic matter. Auxiliary electrode: Platinum. Reference electrode: SCE. Scan rate: 100 mV s-1. A: Anodic start of the CV. Adapted from [10].

On the other hand, according to the literature [10-16], the first stage in the oxidation of dihydroxybenzenes is the formation of a phenoxy-type radical. This radical can be further oxidized to the quinone form or can couple with other radicals or dihydroxybenzenes to form polymers. This formation of polymer may occur in the oxidation of the three isomers, and it can justify the smaller size of the cathodic peak in the oxidation of hydroquinone and catechol (EC mechanism) and also the shift in the oxygen evolution observed in the voltammograms of hydroquinone. However, the polymer formed must be easily removed by the hydroxyl radicals formed during water decomposition as no important differences between successive cycles can be observed for catechol and hydroquinone in the voltammograms of Figure 6. The shift in the hydrogen evolution process toward higher overpotentials has been reported in the literature previously for the reduction of water onto BDD in the presence of aromatics [11-15], but a clear explanation does not exist for this observation.

Avaca et al. [17] studied the behavior of benzene at BDD anodes in acidic solution (0.5 M H_2SO_4) containing 2×10^{-2} M benzene. During the scan toward positive potentials from -1.25 V up to 2.5 V an oxidation peak is observed at 2.0 V (absent in the blank solution), which showed a behavior directly proportional to the scan rate. The observed process at this potential is associated to the irreversible benzene electrochemical oxidation. The linear relationship between the peak current and $v^{1/2}$ obtained by them, it indicates a mass transport control for benzene electrochemical oxidation on BDD. Similar behavior has found for phenol on BDD electrode in acidic medium by Iniesta et al. [18]. However, a different behavior has been observed for the same process on platinum [18] and GC [20]. In one hand, over platinum, the oxidation of benzene produces a peak at ≈1.35 V within the region corresponding to the oxidation of the platinum surface. The oxidation current densities are very small and decrease slightly as the number of cycles is increased. This result indicated that the poisoning of the electrode surface or low activity of the Pt. In contrast, over GC the benzene oxidation depends on events occurring in the oxygen evolution region, but without the presence of a voltammetric peak. In order to confirm the Avaca's group assumptions about that the benzene electrochemical oxidation on BDD is a diffusion controlled process [17], some electrolysis experiments were carried out at constant potential of 2.1 V versus Ag/AgCl at different rotation rates using a BDD rotating disk electrode. An analysis of Figure 7 shows the presence of two other voltammetric peaks, one anodic at 0.7 V and other cathodic at -0.1 V versus Ag/AgCl. During these experiments, they observed the formation of several by-products that are generated through the reaction with (•OH) produced from water electrolysis on the BDD electrode. Some of these intermediates have been identified by other authors under specific experimental conditions, hydroquinone and benzoquinone on GC electrodes [20] and benzoquinone, cyclohexane and CO_2 on platinum [19]. Meanwhile, some of their possible oxidative products such as resorcinol, catechol and hydroquinone have been already studied by Nasr et al. [10], which demonstrated the well defined redox behavior of these intermediates on BDD surface. However, in this study [17], for catechol and hydroquinone the electrochemical behavior was associated to a pair of redox peaks in the same potential region presented in Figure 7b, while for resorcinol only an anodic peak is observed. These features are in a partial agreement with results obtained by Nasr et al., [10], where they justified the presence of a cathodic reduction peak in their voltammograms over BDD electrode and the surface activity for aromatic ring activated with an OH group increases in the ortho- and para-positions (Figures 8 and 9). Thus, hydroquinone and catechol have the entire aromatic ring activated, while carbon five of resorcinol is not activated. This justifies its lower reactivity. Considering that hydroquinone, cathecol and resorcinol are byproducts generated during the benzene electrochemical oxidation over BDD electrodes (see HPLC results in Figure 9) the peaks at 0.7 V and -0.1 V can be attributed to the electrochemical process of these substances. These results differ from those ones for benzene oxidation over GC, where the voltammetric peaks in the region of 0.7 V versus RHE are supposed to be due to the oxidation of hydroquinone and reduction of benzoquinone, respectively, which is a two-electron process.

Furthermore, the benzene oxidation mechanism and its products at BDD (see HPLC results) appear to be different from those at platinum [19]. Thus, in order to confirm the nature of anodic peaks observed in the voltammograms, Figure 8 shows the relationship between the voltametric responses and the benzene concentration. The linear increase of the peak current observed at ≈2.0 V (Figure 8 and inset of the same figure) demonstrated the relationship between the anodic processes observed and the benzene transformation on the electrode.

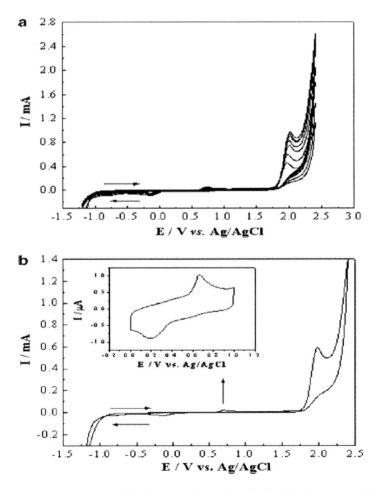

Figure 7. (a) Cyclic voltammograms of the benzene electrochemical oxidation over a BDD electrode in an aqueous sulfuric acid solution 0.5M containing 2×10^{-2} M of benzene at different scan rates: 50, 100, 200, 300, 400, 500 and 600 mV s^{-1} and (b) cyclic voltammogram of the benzene oxidation over a BDD electrode in the same experimental conditions at scan rate = 100 mV s^{-1}. Inset is the potential region between 0 and 1.0 V. Adapted from [17].

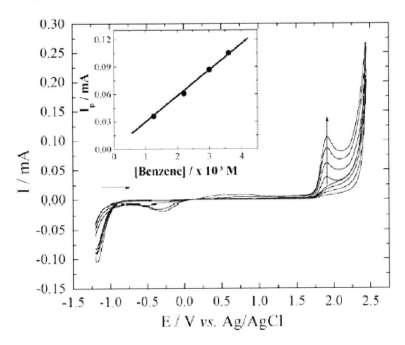

Figure 8. Cyclic voltammograms (first cycle) of the benzene electrochemical oxidation over a BDD electrode in an aqueous sulfuric acid solution 0.5M containing benzene concentrations from 0.36×10^{-3} M to 1.05×10^{-3} M. Scan rate = 100 mV s^{-1}. Inset: Linear relationship between benzene concentration and current peak observed in the same experimental conditions. Adapted from [17].

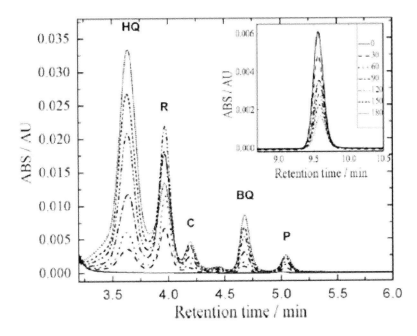

Figure 9. Chromatograms of intermediate products formed during the benzene electro-oxidation (1×10^{-2} M) at 2.4 V versus Ag/AgCl after 180 min in 0.5M H$_2$SO$_4$ supporting electrolyte, products: (HQ) Hydroquinone, (R) Resorcinol, (C) Catechol, (BQ) p-Benzoquinone and (P) Phenol. Inset: Degradation of benzene during the electrolysis. Adapted from [17].

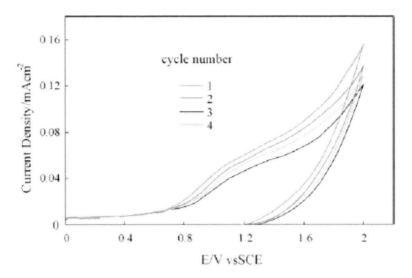

Figure 10. Cyclic voltammograms for the oxidation of BPA at Si/BDD electrode (BPA, 20 mg dm^{-3}; electrolyte, 0.1M Na$_2$SO$_4$; scan rate = 100mVs^{-1}; pH 6; T, 25°C). Adapted from [21].

Anodic oxidation of bisphenol A (BPA) was investigated by Murugananthan et al. [21], and the electro-oxidation behaviour of this organic compound was studied by means of cyclic voltammetric technique as is shown in Figure 10.

In the first cycle, the oxidation peak corresponding to BPA appears approximately at 1.2V vs. SCE. It has been reported that the oxidation peak of BPA at Pt and GC electrode are lower than 0.8V at neutral pH values [22–24]. The difference in oxidation potential could be attributed to the behavior of the electrode material. In the case of BDD, the anodic current peak decreases in the subsequent cycles. This observation can be explained as a result of fouling phenomenon that the polymeric product deposition on the surface of BDD electrode, as reported by other authors [4, 5]. The activity of the electrode surface becomes less due to this passivating behavior. However, this fouling layer can be removed by anodic polarization in the potential region of water decomposition (>2.3 V). This has already been confirmed by some earlier investigations [25, 26] where the activity of BDD surface was restored by anodic polarization method. Moreover, in the range of applied current density in the present study, the anodic potential was found to be in the region of water decomposition thus no electrode fouling problem encountered during the galvanostatic electrolysis. Similar effect was observed by Canizares et al [27] during the oxidation of alcohols and carboxylic acids with diamond electrodes. Therefore, some authors [28-30] have studied the electrochemical behaviour of supporting electrolyte during cyclic voltammetric measurements. Then, Louhichi et al. [28] showed that the nature of supporting electrolyte influences on the efficiency of the electrochemical oxidation of surfactants. Figure 11 shows voltammograms of 0.1M Na$_2$SO$_4$, NaCl and K$_3$PO$_4$ solutions on boron-doped diamond (100mVs^{-1}). As it can be observed, only the voltammogram of the K$_3$PO$_4$ solution presents activity in the potential region of stability of the electrolyte, with two anodic peaks situated at 1.75 and 2.5V (vs. SCE). According to the literature [31, 32], theses peaks can be explained in terms of the oxidation of the phosphates ions (PO$_4^{3-}$, HPO$_4^{2-}$) to phosphates radicals (PO$_4^{2-\bullet}$, HPO$_4^{-\bullet}$), which it is known to be the first step in the formation of the more stable peroxomonophosphate (PO$_5^{3-}$) or peroxodiphosphate (P$_2$O$_8^{4-}$) ions. The oxidation potential of the phosphate radicals increases with their protonation. This explains the two peaks observed in the voltammogram. The voltammograms of the Na$_2$SO$_4$ and NaCl solutions are very similar, although the electrochemical window observed in the sulphate solution voltammogram seems to be slightly wider. This can be explained by the oxidation of chlorides to hypochlorite which overlaps with the oxygen evolution reaction (OER) process and shift the curve towards lower potentials. Although the formation of peroxodisulphates is widely reported in the literature [33] no peaks can be observed in the voltammogram of the sulphate solution. This means that this process should develop in the potential region of water instability.

In the case of cyclic voltammograms with BDD electrodes of solutions containing 2 mM pyrogallol, and Na$_3$PO$_4$ at pH 2 (Figure 12), the voltammograms showed two anodic oxidation peaks which seem to be related to those that appear in the phenol and dihydroxybenzenes voltammograms. An anodic oxidation peak close to the oxygen evolution potential region (2.2-2.4 V vs. SCE) appears. The anodic behavior of pyrogallol at BDD anode is similar than that of dihydroxybenzenes but it is more difficult to be oxidized than them [10]. The oxidation of pyrogallol is accomplished at a lower potential than that of benzoic acid and 2,4,6-trinitrophenol (Table 1). This can be explained by the withdrawing effect of carboxylic and nitro groups which deactivate the aromatic ring. It can be observed also that every peak decreases in size at the second scan. In this range of potential, reverse peaks were not observed. This indicates that these peaks may correspond to irreversible processes, suggesting electrochemical oxidation mechanisms. According to these results, it seems that the anodic oxidation of pyrogallol shares the same initial reaction stages than those of phenolic compounds, as indicated by the presence of related anodic oxidation peaks. This behavior has also been observed in the anodic oxidation of other phenols on BDD electrodes, and it has been previously reported for chlorophenols [12] and nitrophenols [13, 14].

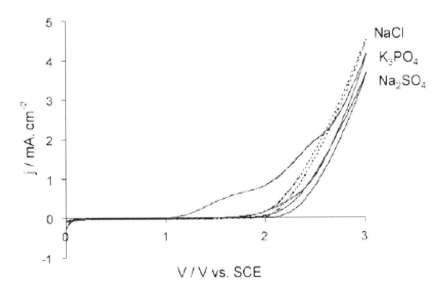

Figure 11. Cyclic voltammograms on BBD anode of three different supporting electrolytes Na$_2$SO$_4$ 0.1M (a), NaCl 0.1M (b) and K$_3$PO$_4$ 0.1M (c). Auxiliary electrode: platinum. Reference electrode: SCE. Scan rate: 100mVs^{-1}. Adapted from [28].

It seems reasonable that these peaks correspond to the oxidation of the phenolic compound to the phenoxy radical and the subsequent oxidation of this phenoxy radical to the corresponding phenoxy cation [4]. It has been reported that the two electrochemically formed compounds are very reactive and can couple to form polymers or undergoes aromatic ring cleavage. On the other hand, the formation of polyhydroxybenzenes is known to be the first step in the aromatic ring cleavage in oxidation mechanisms of phenols. In other works are described the presence of traces of trihydroxybenzenes during the treatment of wastes polluted with phenols [4, 5].

On the other hand, the successive oxidative voltammograms of ethylene glycol, glycerol and diethylene glycol with the scan rate at 100 mV s^{-1} in 0.1 M NaCl solution on BDD showed that these three compounds have the same voltammetric patterns. The first scan, in each voltagramm (Figs. 13a, 13b and 13c), presents one anodic oxidation shoulder which is overlapped by the electrolyte oxidation [30]. It seems that the oxidation of glycols is difficult than that of aromatic compounds [9-15] but similar results were obtained in the cases of aliphatic carboxylic acids and alcohols [9-20]. These results showed that the chemical structure influences largely the voltammetric behavior of organics and can be explained in terms of electronic densities which are higher in aromatics than in aliphatic compounds. This voltammetric study shows that the oxidation of glycols can be carried out directly at BDD anodes in the potential region where the electrolyte is stable.

According to this information, the electrochemical oxidation processes of several model organic compounds can be understood by the cyclic voltammetric behavior observed at BDD surface. These parameters give the features to increase the efficiency of the process, and at the same time, these showed the influence of different factors such as supporting electrolyte, intermediates and organic compound structure.

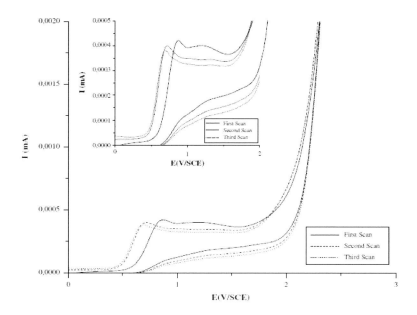

Figure 12. Cyclic voltammograms on BDD anodes of pyrogallol (2 mM) solution on sodium phosphate media (3333.33 mg L^{-1}) at pH=2. Scan rate 100 mV s^{-1}. Auxiliary electrode: platinum. Reference electrode: SCE. (Figure inset: zoom area to observe better the anodic peaks). Adapted from [29].

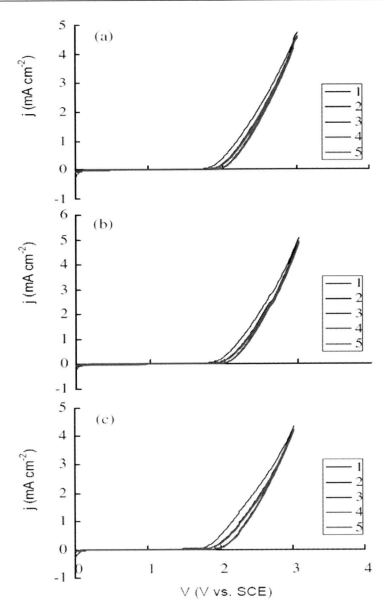

Figure 13. Cyclic voltamogramms on BBD anode of (a) Ethylene glycol 0.5g L^{-1}, (b) Glycerol 0.54 g L^{-1} and (c) Diethylene glycol 0.426 g L^{-1}. Electrolyte: NaCl 0.1 M, Auxiliary electrode: platinum, Reference Electrode: SCE, Scan rate: 100 mV s^{-1}, T = 25 °C. Adapted from [30].

CONCLUDING REMARKS

The above provides an overview of the applications of BDD electrodes to the fields of water treatment for the oxidation of organic pollutants using cyclic voltammetry as an alternative for understanding the electrochemical mechanisms and efficiency of these anodes for removing organic pollutants. Literature results have demonstrated that the complete mineralization of the organics to CO_2 and good Faradic efficiency can be only obtained using high oxygen overpotential anodes, such as boron-doped diamond because these electrodes involve the production of oxygen evolution intermediates, mainly hydroxyl radicals, that non-selectively oxidize the organic pollutants and their intermediates. Despite the numerous advantages of diamond electrodes, they have not yet found wide industrial application, mainly due to their high cost and the difficulties in finding an appropriate substrate on which to deposit the thin diamond layer. In contrast, the electrochemical studies by cyclic voltammetric techniques are a potential tool to prepare new materials or BDD anodes, with low electrical resistivity, good chemical stability and large area, thus being optimum for many applications and in particular for the destruction of water organic pollutants.

ACKNOWLEDGEMENTS

The authors thank to all the authors who gave permission for the reproduction of Figures and Tables. The financial support given for FAPERN (Fundação de Apoio à Pesquisa do Estado do Rio Grande do Norte-Brazil) is acknowledged.

REFERENCES

[1] Chailapakul, O.; Siangproh, W.; Tryk, D. A. *Sensor Lett.* 2006, *4*, 99-119.
[2] Carlisle, J. A. *Nature Materials* 2004, *3*, 668-669.
[3] Patel, B. A.; Bian, X.; Quaiserová-Mocko, V.; Galligan, J. J.; Swain, G. M. *Analyst* 2007, *132*, 41-47.
[4] Panizza, M.; Cerisola, G. *Electrochim. Acta*, 2005, *51*, 191-199.
[5] Martinez-Huitle, C. A.; Ferro, S. *Chem. Soc. Rev.*, 2006, *35*, 1324-1340.
[6] Martínez Huitle, C. A.; Brillas, E. *Appl. Catal. B: Environmental* 2009, *87*, 105-145.
[7] Martínez Huitle, C. A.; Brillas, E. *Angew. Chem. Int. Ed.*, 2008, *47*, 1998-2005.
[8] Pleskov, Y. V.; Sakharova, A. Y.; Krotova, M. D.; Bouilov, L. L.; Spitsyn, B. P. *J. Electroanal. Chem.* 1987, *228*, 19-27.
[9] Kraft, A. *Int. J. Electrochem. Sci.* 2007, *2*, 355–385.
[10] Nasr, B.; Abdellatif, G.; Canizares, P.; Saez, C.; Lobato, J.; Rodrigo. M. A. *Environ. Sci. Technol.* 2005, *39*, 7234-7239.
[11] Canizares, P.; Lobato, J.; Garcıa-Gomez, J.; Rodrigo, M. A. *J. Appl. Electrochem.* 2004, *34*, 111-117.
[12] Canizares, P.; Garcıa-Gomez, J.; Saez, C.; Rodrigo, M. A. *J. Appl. Electrochem.* 2003, *33*, 917-927.
[13] Canizares, P.; Saez, C.; Lobato, J.; Rodrigo. M. A. *Ind. Eng. Chem. Res.* 2004, *43*, 1944-1951.
[14] Canizares, P.; Saez, C.; Lobato, J.; Rodrigo, M. A. *Electrochim. Acta* 2004, *49*, 4641-4650.
[15] Canizares, P.; Saez, C.; Lobato, J.; Rodrigo. M. A. *Ind. Eng. Chem. Res.* 2004, *43*, 6629-6637.
[16] Streitwieser, A.; Heathcock, C. H.; Kosower, E. M. *Introduction to Organic Chemistry*, 4th ed.; MacMillan: New York, 1992.

[17] Oliveira, R. T. S.; Salazar-Banda, G. C.; Santos, M. C.; Calegaro, M. L.; Miwa, D. W.; Machado, S. A. S.; Avaca, L. A. *Chemosphere* 2007, *66*, 2152-2158.

[18] Iniesta, J.; Michaud, P.A.; Panizza, M.; Cerisola, G.; Aldaz, A.; Comninellis, Ch. *Electrochim. Acta* 2001, *46*, 3573–3578.

[19] Montilla, F.; Huerta, F.; Morallon, E.; Vazquez, J.L. *Electrochim. Acta* 2000, *45*, 4271–4277.

[20] Kim, K.-W.; Kuppuswamy, M.; Savinell, R.F. *J. Appl. Electrochem.* 2000, *30*, 543–549.

[21] Murugananthan, M.; Yoshihara, S.; Rakuma, T.; Uehara, N.; Shirakashi, T. *J. Hazard. Mat.* 2008, *154*, 213-220.

[22] Tanaka, S.; Nakata, Y.; Kimura, T.; Yustiawati, Kawasaki, M.; Kuramits, H. *J. Appl. Electrochem.* 2002, *32*, 197–201.

[23] Tanaka, S.; Nakata, Y.; Kuramitz, H.; Kawasaki, M. *Chem. Lett.* 1999, 943–944.

[24] Kuramitz, H.; Matsushita, M.; Tanaka, S. *Water Res.* 2004, *38*, 2331–2338.

[25] Murugananthan, M.; Yoshihara, S.; Rakuma, T.; Uehara, N.; Shirakashi, T. *Electrochim. Acta* 2007, *52*, 3242–3249.

[26] Iniesta, J.; Michaud, P.A.; Panizza, M.; Comninellis, Ch. *Electrochem. Commun.* 2001, *3*, 346–351.

[27] Canizares, P.; Paz, R.; Saez, C.; Rodrigo, M.-A. *Electrochim. Acta* 2008, *53*, 2144-2153.

[28] Louhichi, B.; Ahmadi, M.F.; Bensalah, N.; Gadri, A.; Rodrigo, M.A. *J. Hazard. Mat.* 2008, *158*, 430-437.

[29] Bensalah, N.; Trabelsi, H.; Gadri, A. *J. Environ. Management*, 2009, *90*, 523-530.

[30] Louhichi, B.; Bensalah, N.; Gadri A. *J. Environ. Eng. Manage.* 2008, *18*, 231-237.

[31] Canizares, P.; Larrondo, F.; Lobato, J.; Rodrigo, M.-A.; Saez, C. *J. Electrochem. Soc.* 2005, *152*, D191–D196.

[32] Weiss, E.; Saez, C.; Groenen-Serrano, K.; Canizares, P.; Savall, A.; Rodrigo, M.A. *J. Appl. Electrochem.* 2008, *38*, 93–100.

[33] Michaud, P.-A.; Mahe, E.; Haenni, W.; Perret, A.; Comninellis, Ch. *Electrochem. Solid State Lett.* 2000, *3*, 77.

[34] Louhichi, B.; Bensalash, N.; Gadri, A. *Chem. Eng. Technol.* 2006, *29*, 944-950.

[35] Cañizares, P.; Martínez, F.; Lobato, J.; Rodrigo, M.A. *Ind. Eng. Chem. Res.* 2006, 45, 3474-3480.

In: Diamonds: Properties, Synthesis and Applications
Editors: T. Eisenberg and E. Schreiner, pp. 93-112
ISBN: 978-1-61470-591-8
© 2012 Nova Science Publishers, Inc.

Chapter 5

GRINDING CHARACTERISTICS OF DIAMOND FILM USING COMPOSITE ELECTRO-PLATING IN-PROCESS SHARPENING TECHNIQUE

Yuang-Cherng Chiou, Rong-Tsong Lee and Tai-Jia Chen
Department of Mechanical and Electro-Mechanical Engineering,
National Sun Yat-Sen University, Kaohsiung 80424, Taiwan

ABSTRACT

To finish the CVD diamond film surface, the composite electroplating is simultaneously introduced into the grinding process for sharpening the grinder or disc (briefly called as CEPIS method). In this grinding process, the grinder of cathode and the nickel plate of anode are connected to DC power supply, and they are immersed in plating bath containing diamond particles so that metal ions with diamond particles are deposited onto a grinder in-process to expose fresh sharp particles. Results show that the removal rate of diamond film increases with increasing current density. This removal rate under the current density of 7.5 ASD is 3.8 times higher than that under 0 ASD or the well-known traditional grinding method.

Moreover, the effect of bath composition on the coating structure deposited on a grinder or disc is also investigated. The real area of contact between the grinder and the CVD diamond film increases with increasing nickel chloride concentration due to the influence of the coating structure during the grinding process using a CEPIS method so that the grinder can hold the diamond particles rigidly. Consequently, the grinding ability of the grinder can be significantly improved, where a mirror-like surface of the CVD diamond film can be achieved.

Keywords: Composite Electro-Plating In-process Sharpening (CEPIS), CVD diamond film, Surface roughness, Nickel chloride

1. CONVENTIONAL DIAMOND MACHINING METHODS

Diamonds are in widespread use as precision components in mechanical, thermal, electrical, optical, and electric applications [1], because they provide the excellent physical and chemical properties. However, the surface roughness of the diamond film coated by the CVD process increases with increasing film thickness, and its maximum value can achieve several microns [2]. Hence, the surface roughness must be decreased to meet the actual requirement from sub-microns to nanometers, but the surface machining process is time consuming and expensive [3-12].

The precision machining methods used to finish diamond films include the mechanical, thermo-chemical, chemo-mechanical, ion-beam, and laser methods. In the mechanical polishing method [3-4], the polishing is accomplished by reciprocating the specimen under the diamond wheel. Moreover, a thick polycrystalline CVD diamond film as the polishing abrasive is used to polish another thick CVD diamond film, as shown in Figure 1. The high polishing rate of up to 10 µm/h and surface roughness of the CVD diamond film of 0.55 µm can be achieved using this method.

In the thermo-chemical polishing method, the diamond film is polished using an iron or nickel plate at high temperature (higher than 700°C) so that the carbon atoms on the diamond film can be diffused into the plate in achieving the polishing purpose [5-7]. The schematic drawing for the thermo-chemical polishing apparatus has been shown in Figure 2. The polishing rate of 7µm/h in vacuum at 950°C and the surface roughness, Ra of 2.5 nm of the CVD diamond film are obtained.

In the chemo-mechanical polishing method, the potassium nitrate (KNO_3) or potassium permanganate ($KMnO_4$) is used as oxidizing agent and the ceramic plate or diamond film as polisher, so that the diamond film as specimen is removed at low or medium temperature (about 70-350°C) [8-9]. In this polishing process, the micro/nano cracks are generated on the diamond film due to the action of mechanical force, and the fresh surfaces in the cracks reacted with oxidizing agent should be removed. The schematic drawing for the apparatus of this polishing method has been shown in Figure 3. During the polishing process, the diamond films were coarsely polished against the alumina surface for the polishing time of 3 hours followed by fine polishing against another coarsely polished diamond film for 2 hours. The rms roughness on the polished surfaces of diamond film has decreased from about 0.657 µm to about 0.17 µm.

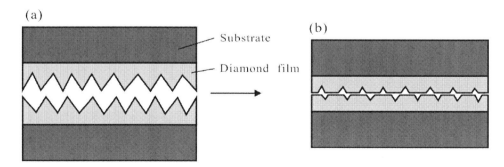

Figure 1. Schematic drawing for the polycrystalline CVD diamond polishing CVD diamond each other : (a) Before polishing and (b) After polishing [4].

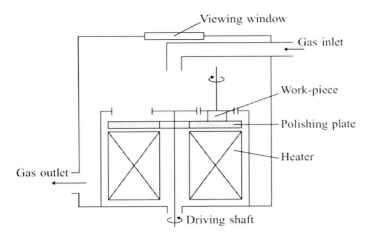

Figure 2. Schematic drawing for the thermo-chemical polishing apparatus [5].

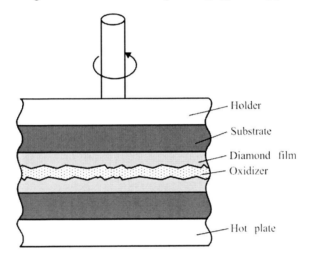

Figure 3. Schematic drawing of the polishing setup using chemo-mechanical technique [8].

In the ion-beam machining methods, the diamond film is removed using ion sources from sputtering [10]. In the laser machining method, the diamond film is evaporated using laser beam [11-12]. Laser and ion beam machining methods are non-contact methods which can be used to finish curved surfaces. However, these machining methods usually need expensive equipments and the machining tools have to be replaced when they were worn out. Therefore, how to promote the removal efficiency of the diamond film has become one of the major topics in the industry and the academia.

2. CEPIS GRINDING METHOD

In the previous study, the authors [13] developed a novel method to lap silicon wafers in which the surface of a lap or disc was composite electroplated so that it held fresh sharp alumina powders during the lapping process.

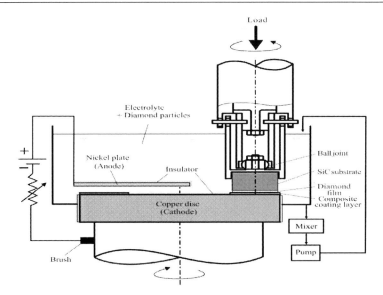

Figure 4. Schematic diagram of the grinding apparatus with composite electroplating in-process sharpening.

The alumina powders were added to the alkaline stannate bath so that a Sn-Al$_2$O$_3$ composite coating layer could be electroplated on the lap. Results showed that the surface of the lap had an unbroken layer with sharp alumina particles and the surface roughness, Ra of the lapped wafer was significantly decreased. It was clear that the worn surface of the lap could be repaired and its lapping capability could be retained during the process. Therefore, it is not necessary to replace and to dress the lap using this method, which we term the composite electro-plating in-process sharpening (CEPIS) method, and its schematic diagram has been shown in Figure 4.

On other hand, composite coatings with different structures can be electroplated using different kinds of co-deposition methods. It is obvious that the coating structure can be significantly influenced by metal ion's concentration of the plating bath. Therefore, the nickel chloride concentration of the plating bath can be adjusted.

The grinding characteristics of CVD diamond film are investigated using CEPIS method, the sharpening mechanism on the grinder or disc is presented, and the nickel chloride concentration of the plating bath is adjusted as follows.

3. CEPIS GRINDING PROCESS USING THE LARGER-SIZED DIAMOND PARTICLES

3.1. Composite Electroplating Process

Before the grinding test, a copper disc is composite electroplated under a current density of 7.5 ASD in 15 minutes for the use on the disc so that its coating thickness is about 40 μm. Its surface features are shown in Figure 5. Figures 5(a) and (b) depict the secondary electron image (SEI) and backscattered electron image (BEI) micrographs of the coating surface, respectively.

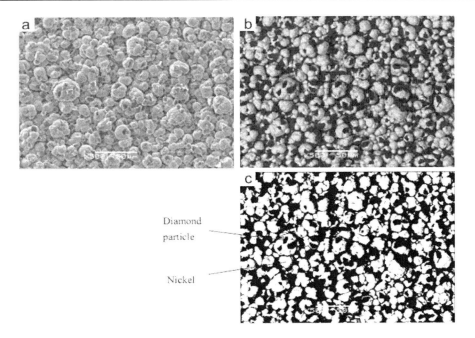

Figure 5. The SEM micrographs for Ni-Diamond composite coating on the disc after electroplating 15 min: (a) SEI, (b) BEI, and (c) Binarization.

Figure 5(a) shows that the composite coating structure is porous so that the diamond particles with 10 μm in size suspended in the bath can be retained on the coating surface. Figure 5(b) shows that the diamond particles (black dots) are uniformly distributed over the nickel metal (gray color). To calculate the area covered by the diamond particles, Figure 5(b) can be transformed into a BEI binarization using image analysis, as shown in Figure 5(c). In this figure, the black dots are diamond particles and the white area represents nickel phase. The area covered by the diamond particles can be measured by calculating the black dots of the image. In this figure, the percentage covered by the diamond particles is about 50%.

3.2. Coating Thickness on the Disc and Removal Amount of CVD Diamond Film

Figure 6 shows the coating thickness on the disc or grinder during the grinding process under a load of 35 N, a work-piece speed of 200 rpm, and a disc speed of 200 rpm. Figure 6 shows that the coating thickness on the disc varies linearly with grinding time at different current densities. It is clear that the coating thickness on the disc decreases from 40 μm to 30 μm in 30 min at 0 ASD, because the coating layer is worn out during the grinding process. However, the coating thickness increases from 40 μm to 51 μm under 2.5 ASD, and from 40 μm to 71 μm 7.5 ASD due to the generation of a new coating layer.

Figure 7 shows the removal amount of the diamond film during the grinding process using the same operating parameters as Figure 6. Figure 7 shows that the removal amount of the diamond film increases linearly with the grinding time at different current densities. The amount of the diamond film removed in 30 min is 0.7 mg at 0 ASD, 1.2 mg at 2.5 ASD, and 2.4 mg at 7.5 ASD, respectively.

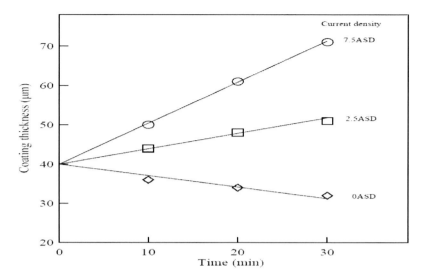

Figure 6. Variation for the coating thickness on the disc under different current densities in grinding process.

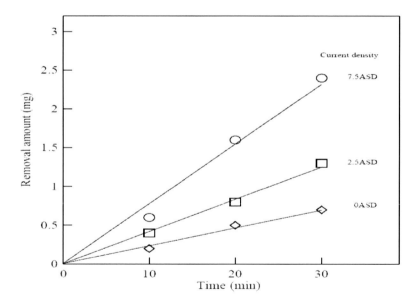

Figure 7. Variation for the removal amount of diamond film under different current densities in the grinding process.

The coating thickness on the disc and the removal amount of the diamond film are linearly proportional to the grinding time, as shown in Figs. 6 and 7. These two figures makes clear that the removal rate of the diamond film is related to the variation rate of the coating thickness on the disc. To investigate their relationship, the variation rate of the coating thickness and the removal rate of the diamond film under different current densities are depicted in Figure 8. It shows that the variation rate of the coating thickness is -0.267μm/min and the removal rate of the diamond film is 0.023mg/min at 0 ASD, 0.367μm/min and 0.043mg/min at 2.5 ASD, and 1.03μm/min and 0.08mg/min at 7.5 ASD, respectively.

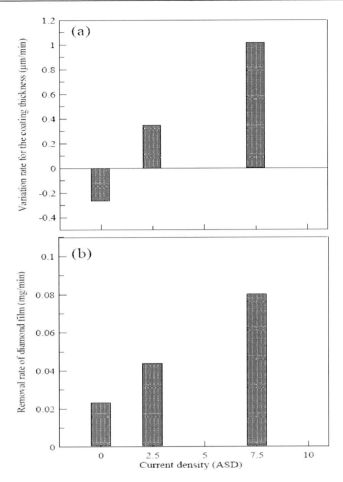

Figure 8. Variation rate of the coating thickness on the disc and removal rate of diamond film under different current densities in the grinding process.

Based on the results shown in Figs. 6 to 8, it is clear that the coating layer on the disc is worn away at 0 ASD since its coating thickness decreases. Hence, the disc or grinder has to be replaced periodically. Furthermore, its removal rate of diamond film is also the lowest among all current densities. On the other hand, when the current density increases from 0 ASD to 2.5 ASD, although the coating layer on the disc is still worn away, it can be repaired with composite electroplating, which replenishes the coating thickness. In addition, the removal rate of diamond film is about 1.7 times higher than that of the traditional grinding method. This indicates that the coating surface on the disc can be sharpened using the CEPIS method. At 7.5 ASD, the coating thickness on the disc is rapidly increased, and the removal rate of the diamond film achieves its highest value, which is about 3.8 times higher than that of the traditional grinding method. This indicates that the effect of the current density on the sharpening disc or grinder is significant. The sharpness of the coating surface on the disc increases with increasing current density, in turn increasing the removal rate of the diamond film.

The experimental results show that the coating thickness on the disc can be maintained and the sharpness on the disc surface can be renewed by using CEPIS method, leading to higher removal rate of the diamond film during the grinding process.

3.3. Variation of Surface Roughness of CVD Diamond Film during Grinding Process

To understand the grinding efficiency of the diamond film using the CEPIS method, a surface profilometer is used to measure the profile and the surface roughness of the diamond film, as shown in Figs. 9 and 10, respectively. Figure 9 shows that the profile of the diamond film in the early grinding stage at 7.5 ASD is flatter than that at 0 ASD.

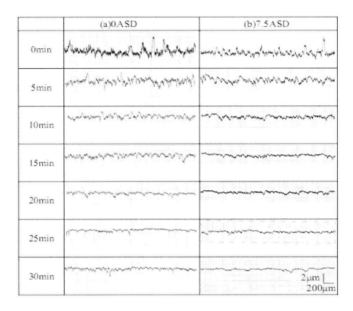

Figure 9. Variation for the profile of diamond film in the grinding process under different current densities: (a) 0 ASD, (b) 7.5 ASD.

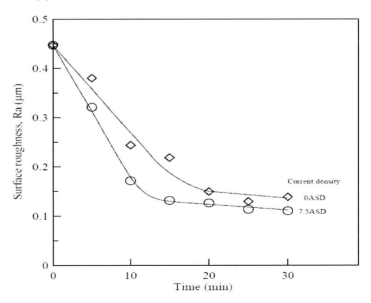

Figure 10. Variation for the surface roughness of diamond film in grinding process under different current densities.

Figure 10 makes clear that the surface roughness of the diamond film decreases rapidly with increasing grinding time, and Ra approaches to 0.13 μm in 15 min at 7.5 ASD, and in 25 min at 0 ASD. This indicates that the surface roughness of the diamond film can be reduced to a smaller value in a shorter grinding time using the CEPIS method.

3.4. Sharpening Mechanism of the Disc or Grinder

In light of the aforementioned grinding results of diamond film, the removal rate of the diamond film increases, and the surface roughness of diamond film decreases using the CEPIS method. These indicate that the disc surface is sharpened by introducing the composite electroplating process. In order to investigate the sharpening mechanism of the disc surface, the SEM is used to observe the structure of coating layer on the disc at corresponding locations. Figs. 11 and 12 show the typical SEM micrographs under different current densities.

Figures 11 and 12 illustrate the three corresponding locations on the surface of the composite coating layer at 0 ASD and 7.5 ASD, respectively. Figure 11(a) shows that the marked areas A, B, and C exhibit porous features after a grinding time of 20 min, but these three areas become flatter after a grinding time of 30 min, as shown in Figure 11(b). Further, the diamond particles in the bath are embedded in the Ni base (see the marked areas A', B', and C'). These transitions of surface result from the decrease of coating thickness during the grinding process (Figure 6). In the same way, Fig 12 shows the marked areas A, B, and C and A', B', and C' for grinding times of 20 min and 30 min, respectively. In these areas, the surface features are not significantly changed, and the distribution of the diamond particles on the coating surface remains during the grinding process from 20 min (Figure 12(a)) to 30 min (Figure 12(b)). This indicates that sharpening condition is continuously occurring on the surface of the disc by the increase of the coating thickness during the grinding process (Figure 6).

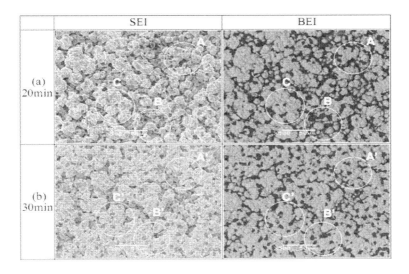

Figure 11. SEM micrographs of the composite coating surface on the disc under 0 ASD after the grinding time of: (a) 20 min, (b) 30 min.

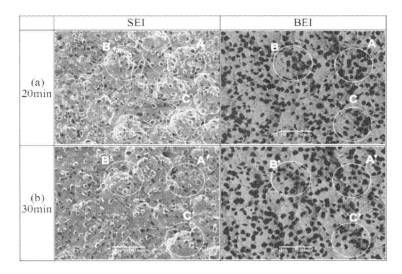

Figure 12. SEM micrographs of the composite coating surface on the disc at 7.5 ASD after a grinding time of: (a) 20 min, and (b) 30 min.

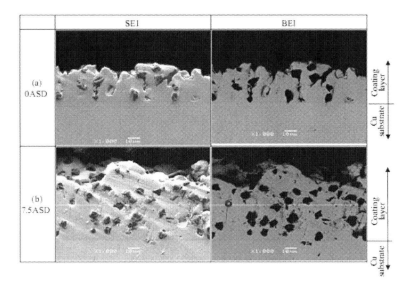

Figure 13. SEM micrographs of the cross-section of a coated layer on the disc after a grinding time of 30 min at different current densities: (a) 0 ASD, and (b) 7.5 ASD.

This experiment not only observes the surface of the composite coating layer (Figs. 11 and 12), but also the cross sections of a coated layer on the disc under different current densities, as shown in Figs. 13(a) and (b). It is clear that the coating layer is worn away, a fewer diamond particles exposed on the coating surface at 0 ASD during the grinding process, as shown in Figure 13(a). However, Figure 13(b) shows that the coating layer is continuously renewed so that many diamond particles are uniformly distributed on the coating layer at 7.5 ASD. Consequently, the sharpening condition takes place during the grinding process.

The sharpening mechanism of the disc during the CEPIS grinding process can be deduced from Figs.12 and 13(b), and it is shown in Figure 14. To explain in detail the sharpening mechanism of the disc under a given current density, it is necessary to understand the contact condition between the diamond film and the disc or grinder in the grinding apparatus shown in Figure 1. Although the diamond film is always in contact with the disc surface, only about 5% of the disc surface area is carrying out the grinding during each rotation of the disc. Hence, it is obvious that when this disc surface area is in contact with the diamond film, the coating thickness and the number of diamond particles decreases as in the traditional grinding method, as shown in Figs. 14(a) and 14(b). However, about 95% of the disc surface area is immersed in the bath so that the coating thickness and the number of diamond particles are patched up during each rotation of the disc, as shown in Figs. 14(b) and 14(c). During this period, the active grit density is increased so that the removal rate can be increased when this area on the disc enters the grinding zone. Moreover, the active grit density increases with increasing current density so that the removal rate at 2.5 ASD is lower than at 7.5 ASD. Consequently, the grinding capability is increased by the CEPIS method and new diamond particles are continuously exposed on the disc surface. Moreover, the coating thickness and the number of diamond particles exposed on the disc surface can be controlled by adjusting the current density and the operating parameters. The disc or grinder can retain its sharpness, lengthening its life considerably. Thus, the Composite Electro-Plating In-process Sharpening (CEPIS) method offers high efficiency with low cost.

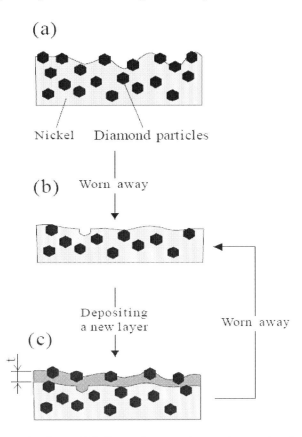

Figure 14. Sharpening mechanism of Ni-diamond composite coating on the disc during the CEPIS grinding process: (a) original, (b) worn away, and (c) depositing a new layer during each rotation on the disc, where t is the incremental amount of the coating in thickness, and its actual value is much smaller than shown here (t is about 0.05 μm/rev at 7.5 ASD).

4. CEPIS Grinding Characteristics under Different Nickel Chloride Concentrations

4.1. Coating Thickness and Coating Structure on the Disc during the Electroplating Process

In this section, the grinding characteristics of CVD diamond film are investigated using CEPIS method under different nickel chloride concentrations in the plating bath. Moreover, the smaller size of diamond particles, 3μm, mixed the plating bath is also employed. The coating structure on the disc surface is significantly influenced by nickel chloride concentration during the electroplating process. Hence, it is necessary to investigate the deposited characteristics of the composite coating under different nickel chloride concentrations in the bath. Figure 15 shows the typical results of the coating thickness histories on the disc surface during the electroplating process. The operating parameters for this case are set to a current density of 7.5 ASD and a disc speed of 50 rpm. In Figure 15, it clearly shows that the coating thickness on the disc surface increases linearly with time under different nickel chloride concentrations. After the electroplating time of 30 minutes, the coating thickness on the disc surface increases to 77 μm at the nickel chloride concentration of 10 g/L, to 51 μm at the 30 g/L, and to 38 μm at the 75 g/L, respectively.

In order to understand the relationship between the growth rate of the coating and the nickel chloride concentration, the above results are rearranged and plotted in Figure 16. Figure 16 shows that the growth rate of the coating decreases with increasing nickel chloride concentration. When the concentration is increased from 10 to 75 g/L, the growth rate of the coating is decreased from 2.56 to 1.26μm/min.

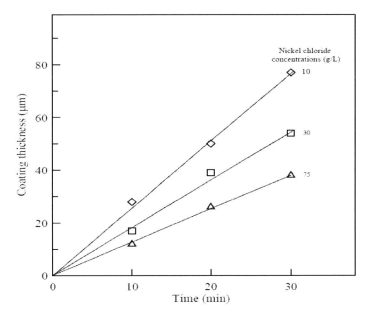

Figure 15. Variation for the coating thickness on the disc during the electroplating process at a current density of 7.5 ASD, the disc speed of 50 rpm and the electroplating time of 30 min under different nickel chloride concentrations.

Grinding Characteristics of Diamond Film Using Composite Electro-Plating... 105

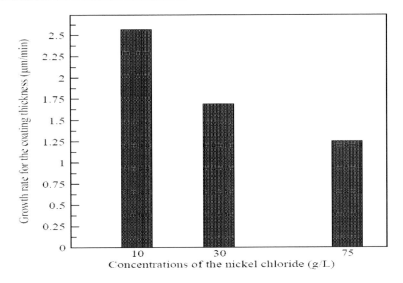

Figure 16. Growth rate for the coating thickness on the disc during the electroplating process at a current density of 7.5 ASD, the disc speed of 50 rpm and the electroplating time of 30 min under different nickel chloride concentrations.

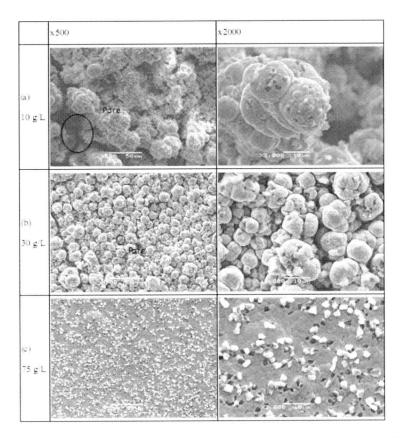

Figure 17. SEM micrographs for Ni-Diamond composite coating on the disc at a current density of 7.5 ASD, the disc speed of 50 rpm after electroplating 30 min under different nickel chloride concentrations: (a) 10 g/L, (b) 30 g/L, and (c) 75 g/L.

To investigate this phenomenon, the coating surface of the disc is examined using a scanning electron microscope (SEM). Figure 17 shows the SEM micrographs for Ni-Diamond composite coating on the disc after the electroplating time of 30 min under different nickel chloride concentrations. Figure 17(a) shows that deposits with a "cauliflower" structure are observed, and they are formed by microcrystalline aggregates at low nickel chloride concentration. However, the size of crystals are decreased with increasing nickel chloride concentration, so that "ball-like" agglomerates (Figure 17(b)), fine-grains and compact deposits (Figure 17(c)) can be observed. Furthermore, these agglomerates are characterized with the presence of pores or valley areas.

The average pore diameter of the coating is about 50μm at the nickel chloride concentration of 10 g/L, 12 μm at 30 g/L, respectively, and almost no pores can be detected at 75 g/L. It makes clear that the composite coating is loosely stacked at lower concentration so that the coating thickness becomes thicker. Hence, the corresponding coatings layers have a poor mechanical property. When the nickel chloride concentration is increased to 75 g/L, the corresponding coatings have an excellent mechanical property with lower growth rate. It is evident that the nickel chloride concentration plays an important role to form compact coating layers at high current density. In other words, an increase in the nickel chloride concentration conducts that the coating layer has a good wear resistance. These results are quite similar to those obtained by Marozzi and Chialvo [16] with NH_4Cl concentration.

4.2. Coating Thickness on the Disc during the Grinding Process

After the electroplating time of 30 min on the surface on the disc, a CVD diamond film surface is ground using the CEPIS method. Figure 18 shows the typical results of the coating thickness histories on the disc during the grinding process. The operating parameters are a load of 35 N, a disc speed of 50 rpm, and work-piece speed of 400 rpm. In Figure 18, it is clear that the coating thickness on the disc is decreased with increasing grinding time under the current density of 0 ASD (or traditional grinding method), but it is increased with increasing grinding time under the current density of 2.5 and 7.5 ASD (CEPIS method).

The increment or decrement amount of the coating thickness on the disc is linearly proportional to grinding time under different nickel chloride concentrations and current densities, as shown in Figure 18. Therefore, the average growth rate or the decrement rate of the coating thickness on the disc under different nickel chloride concentrations can be depicted as Figure 19. Figure 19 shows that when the nickel chloride concentrations are increased from 10 to 75 g/L, the decrement rate of the coating thickness is changed from -0.38 μm/min to -0.17μm/min under 0 ASD, but the growth rate is changed from 2 μm/min to 1.1μm/min under 7.5 ASD. These results indicate that the looser structure of the coating has larger wear rate during the grinding process under the traditional grinding method (0 ASD). However, the coating thickness on the disc (Figure 18) continues to show signs of improvement during the grinding process under different nickel chloride concentrations at 2.5 and 7.5 ASD. This indicates that although the composite coating on the disc is worn out during the grinding process, it still can be simultaneously repaired using CEPIS method. Figure 19 shows that the increment rate of the coating thickness at the 7.5 ASD is larger than that at the 2.5 ASD. Furthermore, it is also given from Figs. 3 and 4 that the coating structure on the disc during the electroplating process varies from looser to denser with increasing nickel chloride concentration. Hence, the increment rate of the coating thickness is gradually decreased with increasing nickel chloride concentration, as shown in Figure 19.

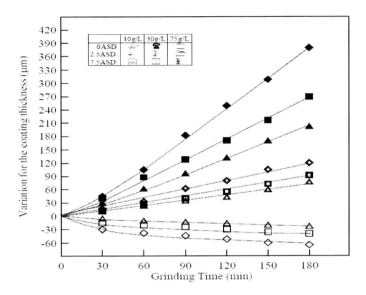

Figure 18. Variation for the coating thickness on the disc during the grinding process at the disc speed of 50 rpm, the work-piece speed of 400 rpm, and the grinding time of 180 min under different nickel chloride concentrations and different current densities.

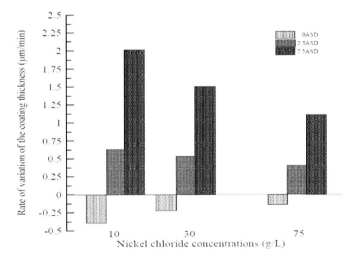

Figure 19. Rate of the variation of the coating thickness on the disc during the grinding process at the disc speed of 50 rpm, the work-piece speed of 400 rpm, and the grinding time of 180 min under different nickel chloride concentrations and current densities.

4.3. Surface Roughness of CVD Diamond Film during Grinding Process

To understand the grinding characteristics of the diamond film under different nickel chloride concentrations using CEPIS method, the surface profilometer is used to measure the surface roughness of the diamond film after the grinding process, as shown in Figure 20. Figure 20 shows the surface roughness, Ra of the diamond film versus the grinding time under a load of 35 N, a lower spindle of 50 rpm, an upper spindle of 400 rpm, and the grinding time of 180 min. This figure makes clear that when the nickel chloride concentration is increased from 10 to 75 g/L, the surface roughness is decreased from 0.17 to 0.08 μm under 0 ASD. On the other hand, the surface roughness is decreased with grinding time at constant current density and with increasing current density at constant grinding time. Generally speaking, the higher current density with higher nickel chloride concentration can achieve a good grinding ability using CEPIS method in the present cases. As the nickel chloride concentration is increased from 10 to 75 g/L, the surface roughness of the diamond film decreases from 0.148 to 0.065 μm at the 2.5 ASD, and from 0.14 to 0.03 μm at 7.5 ASD. Figure 21 shows the surface profile histories of the diamond film at different nickel chloride concentrations during the grinding process under 7.5 ASD using the CEPIS method. In Figure 21(a), it is clear that the density of peaks on the surface profile of diamond film almost remains constant with smaller amplitude after the grinding time of 180 min at the nickel chloride concentration of 10 g/L. However, in Figure 21(b), a certain amount of the peaks has been removed after the grinding time of 120 min at the nickel chloride concentration of 75 g/L. These results indicate that the higher the nickel chloride concentration has the better the grinding ability using the CEPIS method.

In addition, the surface of the diamond film is observed using SEM under the nickel chloride concentration of 75 g/L after the grinding time of 180 min, as shown in Figure 22. Figure 22 shows the peaks on the surface of the diamond film have been removed by the grinding. This indicates that the peaks of the surface of the diamond film can be rapidly removed using the CEPIS method.

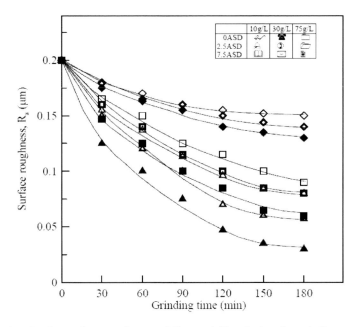

Figure 20. Variation for the surface roughness of diamond film during the grinding process at the disc speed of 50 rpm, the work-piece speed of 400 rpm, and the grinding time of 180 min under different nickel chloride concentrations and current densities.

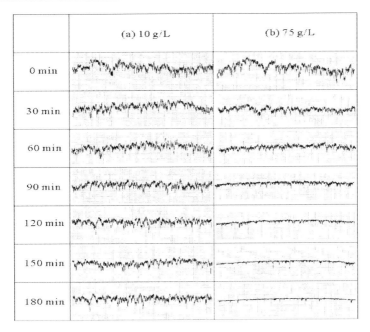

Figure 21. Surface profile histories of the diamond film within the grinding time of 180 min at the disc speed of 50 rpm, the work-piece speed of 400 rpm, the current density of 7.5 ASD under different nickel chloride concentrations: (a) 10 and (b) 75 g/L.

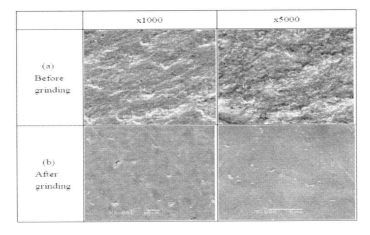

Figure 22. SEM micrographs of the CVD diamond film after the grinding time of 180 min under 7.5 ASD at the nickel chloride concentration of 75 g/L.

4.4. Sharpening Mechanism on the Disc under Different Nickel Chloride Concentrations

As mentioned above, the rough surface profile of diamond film can be polished during the grinding process using the CEPIS method in which the grinder is simultaneously sharpened by composite electroplating. Since the structure of coating layer on the disc or grinder is significantly influenced by the current density and the nickel chloride concentration, it is necessary to observe the disc surface using SEM. Figs. 23 shows the SEM micrographs of the Ni-diamond composite coating on the disc surface after the grinding time of 180 min under different nickel chloride concentrations at 0 ASD and 7.5 ASD, respectively. In Figure 23(1), it makes clear that their coating structures on the disc are almost the same at 0 ASD due to no composite electroplating during the grinding process. On the other hand, in Figure 23(2), it makes clear that their coating structures on the disc are quite different at 7.5 ASD.

Generally, the coating structure on the disc becomes from looser to denser and the real area of contact between the grinder and the diamond film with increasing nickel chloride concentrations from 10 to 75 g/L.

In the CPEIS process, the real area of contact between the grinder and the diamond film is about 15~25 % of the apparent area of contact under the nickel chloride concentration of 10 g/L so that the renewed diamond particles may be exposed on the valley area on the disc surface profile (see the marked area A). Therefore, only few diamond particles on the coating surface become active in grinding the diamond film so that the decrement for the surface roughness of the diamond film is poorer than that using traditional method. As nickel chloride concentration is increased to 30 g/L, the real area of contact is increased to about 45~55 % of the apparent area of contact so that more diamond particles embedded on the coating surface become active in grinding the diamond film, and the grinding ability on the disc is increased. As the nickel chloride concentration increases to 75 g/L, it becomes very smooth without porous area so that the diamond particles embedded on the coating surface is very active in grinding the diamond film (see the marked area B). Since the nickel substrate can hold the diamond particles rigidly, the mirror-like surface of the CVD diamond film (Ra=0.03μm) can be achieved.

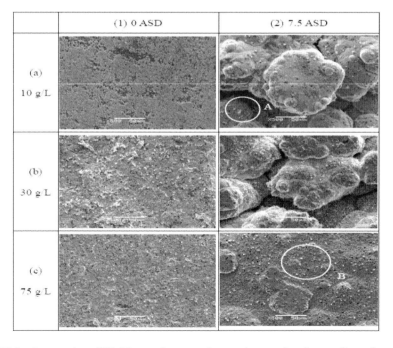

Figure 23. SEM micrographs of Ni-Diamond composite coating on the disc surface after the grinding time of 180 min under different current densities: (1) 0 ASD, (2) 7.5 ASD.

CONCLUSION

The Composite Electro-Plating In-process Sharpening (CEPIS) method for grinding is proposed to finish CVD diamond film. In this method, the current density is applied between the disc (cathode) and the nickel plate (anode) to conduct composite electroplating so that it is not necessary to replace the grinder during the grinding process. The influences of the current densities and bath composition on coating structure deposited on the disc and the grinding characteristics of the diamond film are investigated. Based on the experimental results and the SEM observations on the coated disc and the diamond film, the main results are outlined below:

1. The removal rate of the diamond film increases with increasing current density, and the highest removal rate is about 3.8 times higher than that of the traditional grinding method at the particle size of 10 μm. Meanwhile, the variation rate of the coating thickness on the disc goes from negative to positive as current density increases from 0 ASD to 7.5ASD.
2. During the grinding process, the surface roughness decreases with increasing current density and nickel chloride concentration at constant grinding time. The higher current density and the higher nickel chloride concentration, the better the grinding ability using the CEPIS method in the present cases.
3. The real area of contact between the grinder and the diamond film increases with increasing nickel chloride concentration at constant current density. The smaller the real area of contact, the smaller amount of the renewed diamond particles is exposed on the valley area on the disc surface so that only few diamond particles on the coating surface become active in grinding the diamond film. With the larger real area of contact, the diamond particles are embedded on the coating so that they become very active in grinding the diamond film, and the mirror-like surface of the CVD diamond film (Ra=0.03μm) can be achieved.
4. The sharpening mechanism for the CEPIS method is deduced. This mechanism indicates that the coating thickness on the disc increases with grinding time using the CEPIS method, enabling higher grinding capabilities because a larger number of diamond particles exposed on the coating surface on the disc is maintained.

REFERENCES

[1] A. Gicquel, K. Hassouni, F. Silva, and J. Achard, CVD diamond films: from growth to applications, *Current Applied Physics*, 1 (2001) 479-496.

[2] J. E. Graebner, S. Jin, G. W. Kamlott, J.A. Herb, and C.F. Gardinier, Unusually high thermal conductivity in diamond films, *Applied Physics Letter*, 60 (1992) 1576-1578.

[3] S. E. Grillo, J. E. Field, and F. M. van Bouwelen, Diamond polishing: the dependency of friction and wear on load and crystal orientation, *Journal of Physics D: Applied Physics*, 33 (2000) 985-990.

[4] C. J. Tang, A. J. Neves, A. J. S. Fernandes, and J. Grácio, N. Ali, A new elegant technique for polishing CVD diamond films, *Diamond and Related Materials*, 12 (2003) 1411-1416.

[5] H. Tokura, C. F. Yang, and M. Yoshikawa, Study on the polishing of chemically vapour deposited diamond film, *Thin Solid Films*, 212 (1992) 49-55.

[6] A. M. Zaitsev, G.. Kosaka, V. Raiko, R. Job, T. Fries, and W. R. Fahrner, Thermochemical polishing of CVD diamond films, *Diamond and Related Materials*, 7 (1998) 1108-1117.

[7] R. Ramesham, and M. F Rose, Polishing of polycrystalline diamond by hot nickel surface, *Thin Solid Films*, 320, (1998) 223-227.

[8] B. Bhushan, V. V. Subramaniam, A. Malshe, B. K. Gupta, and J. Ruan, Tribological properties of polished diamond films, *Journal of Applied Physics*, 74 (1993) 4174-4180.

[9] C.Y. Cheng, H.Y. Tsai, C.H. Wu, P.Y. Liu, C.H. Hsieh, and Y.Y. Chang, An oxidation enhanced mechanical polishing technique for CVD diamond films, *Diamond and Related Materials*, 14 (2005) 622-625.

[10] A. Hirata, H. Tokura, and M. Yoshikawa, Smoothing of chemically vapour deposited diamond films by ion beam irradiation, *Thin Solid Films*, 212 (1992) 43-48.

[11] S. M. Pimenov, A. A. Smolin, and V. I. Konov, Smoothening of diamond films with an ArF laser, *Diamond and Related Materials*, 1 (1992) 782-788.

[12] V. N. Tokarev, J. I. B. Wilson, M. G Jubber, P. John, and D. K. Milne, Modeling of self-limiting laser ablation of rough surfaces: application to the polishing of diamond films, *Diamond and Related Materials*, 4 (1995) 169-176.

[13] Y. C. Chiou, R. T. Lee and C. L. Yau, A novel method of composite electroplating on lap in lapping process, *International Journal of Machine Tools and Manufacture*, 47 (2007) 361-367.

[14] T. J. Chen, Y. C. Chiou, and R. T. Lee, Grinding Characteristics of Diamond Film Using Composite Electroplating In-process Sharpening Method, *International Journal of Machine Tools and Manufacture*, 49 (2009) 470-477.

[15] T. J. Chen, R. T. Lee, and Y. C. Chiou, Sharpening Mechanism Using Composite Electro-Plating In-Process Sharpening Technique, *Advanced Materials Research*, 126-128 (2010) 633-638.

[16] C. A. Marozzi and A. C. Chialvo, Development of electrode morphologies of interest in electrocatalysis. Part1: Electrodeposited porous nickel electrodes, *Electrochimica Acta*, 45 (2000) 2111-2120.

[17] A. P. Malshe, B. S. Park, W. D. Brown, and H. A. Naseem, A review of techniques for polishing and planarizing chemically vapor-deposited (CVD) diamond films and substrates, *Diamond and Related Materials* 8 (1999) 1198-1213.

In: Diamonds: Properties, Synthesis and Applications ISBN: 978-1-61470-591-8
Editors: T. Eisenberg and E. Schreiner, pp. 113-130 © 2012 Nova Science Publishers, Inc.

Chapter 6

DIAMOND DEPOSITED BY A DIRECT CURRENT HOT CATHODE PLASMA DISCHARGE: CHARACTERIZATION AND APPLICATIONS

M. Reinoso,[1,2,3] *F. Álvarez,*[1,2] *E. B. Halac,*[1,2] *and H. Huck*[1,2]

[1]Departmento de Física Experimental. Comisión Nacional de Energía Atómica, Av. General Paz 1499 (1650). General San Martín - Buenos Aires, Argentina
[2]Niversidad Nacional de San Martín, Martín de Irigoyen N° 3100 (B1650), Villa Ballester- Buenos Aires, Argentina
[3]Onsejo Nacional de Investigaciones Científicas y Técnicas, Rivadavia 1917, (1033) Buenos Aires, Argentina

ABSTRACT

A diamond deposition process is introduced and discussed together with its applications to several substrates.

Diamond films are deposited by a direct current plasma discharge, assisted by electron emission from a hot cathode consisting of one or more tungsten filaments. This method allows good quality diamond film growth at a rate of up to 10 μm h^{-1} over a relatively large area (8 cm^2). This deposition method is evaluated on different substrates (diamond, silicon, steels and alumina). Several surface pre-treatments are employed for each substrate: silicon diffusion, silicon interlayers, diamond seeding, and surface scratching using diamond and SiC powder. The results are described and analyzed in detail in each case.

Deposited diamonds are characterized by Raman spectroscopy, electron microscopy (SEM), energy dispersive spectroscopy (EDS), and X-ray diffraction (XRD). Potential applications are described below.

1. INTRODUCTION

Since ancient times, diamonds have been regarded as a bewitching element. They have been an emblem of authority, status, love, and magical power [1]. During the twentieth century, the capability of producing diamond in laboratories has opened up a world of possibilities for industrial and technological applications in different fields such as electronic, mechanics, or biology. Subsequently, the increasing interest has impelled plentiful investigations on the subject.

The outstanding physical and chemical properties of diamond (hardness, transparency, inertness, etc.) have made diamond coatings an appealing subject of research for a wide spectrum of applications: protective coatings in chemically aggressive environments, laser windows, etc. Many different CVD deposition techniques have proved satisfactory for diamond film growth, and good quality diamond films have been obtained even by means of relatively low cost technology such as Hot Filament Chemical Vapor Deposition (HFCVD) [2]. However, unlike many hard coating materials with widespread applications, major drawbacks in the quality of deposited diamond films are strongly related in many cases to substrate composition and surface treatments performed prior to deposition.

In the present chapter, a deposition method is described consisting of a hot tungsten filament acting as a hot cathode in a direct current plasma discharge. A brief description and characterization of diamond films deposited on silicon wafers is presented to illustrate the deposition method. Subsequently, diamond deposition is studied on two different substrates: stainless steel and alumina that respectively require pretreatment of the surface and limiting film thickness to achieve good quality adherent diamond film growth. Also, thicker diamond films deposited on specifically prepared alumina samples can be detached from the substrate to obtain self-supporting films. Finally, a review of present and potential applications of self-supporting diamond films is presented.

2. DEPOSITION AND CHARACTERIZATION

In the growth method presented here, diamond deposition is carried out in a 15×10^3 cm^3 stainless steel chamber. After loading the samples, the chamber is evacuated to a pressure of 10^{-6} mbar by means of a turbomolecular pump. The vacuum system is then switched to a roughing line consisting of a mechanical pump which is used during the deposition process that takes place at a pressure of 20 mbar.

A 0.9 mm diameter hot tungsten filament (6V, 40A), which acts as a hot cathode, is used as an electron emitter, and it is positioned 2.5 cm away from the sample holder. A voltage of about 200 V is initially applied between the filament and the sample holder in order to start the plasma.

The output current is limited so that current densities at the surface of the sample holder are in the range of 0.3 to 0.5 A cm^{-2} so that, after plasma is initiated, a final voltage drop of about 150 V is produced between the filament and the sample holder. This allows temperatures in the range of 800 to 1,000 °C to be achieved on the sample holder. No extra heating system is used for the samples. The temperature of the tungsten filament is 2,300 °C during the process, as measured by an optical pyrometer.

A total gas flow of 500 sccm consisting of methane and hydrogen is injected into the chamber. Methane and hydrogen are injected separately at a ratio of 0.002 CH$_4$:H$_2$ from two different nozzles, as can be seen in Figure 1.

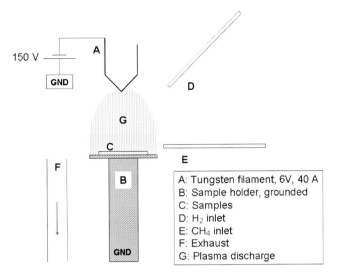

Figure 1. Schematic diagram of the deposition system.

The exhaust is located at the same height as the sample holder. Methane is injected a few mm above and at 1 cm horizontally from the opposite side of the sample holder so that it flows tangentially to the sample surface into the exhaust. Hydrogen is injected at 1 cm above the samples, and horizontally at 1 cm from the sample holder, as can be seen in the diagram of Figure 2. This geometry is settled to minimize methane concentration in the hotter region close to the filament because in this way, a lower content of non-diamond carbon deposition is observed in the resulting film.

A plasma discharge is thus formed, covering an area of about 5 to 6 cm^2 on the sample holder (anode), where a uniform diamond film is deposited.

For larger areas, this system is replicated with the addition of another filament and two extra inlets corresponding to methane and hydrogen. The exhaust is not duplicated.

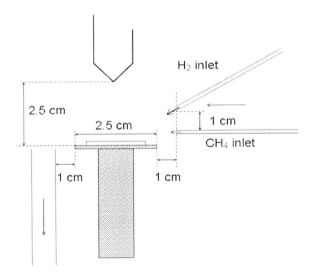

Figure 2. Diagram showing the geometry of inlet gases, filament and sample holder positions in the growing chamber.

Diamond coatings on silicon wafers are studied to illustrate the performance of the deposition apparatus. Results of diamond deposition on stainless steel (AISI 304 and AISI 316) and three different alumina substrates are presented. Diamond films can be deposited by this method at high growth rates of up to 10 μm h^{-1} on silicon and alumina substrates and up to 3 μm h^{-1} on stainless steel.

The microstructure of the films is studied using X-ray diffraction (XRD) and Raman spectroscopy. Grazing-incidence X-ray diffraction is employed to study the crystalline structure of the films (incidence angle: 3 °). The XRD patterns are recorded over the 2θ range of 10–90 °, using Cu Kα radiation (1.54184 Å) at room temperature (scanning step 0.02 °, 1 s). Raman spectroscopy is the usual technique to characterize carbon deposits. It gives bulk information about the bonding character of the atoms (i.e. coordination and bond-angle disorder). Raman spectra are recorded using a LabRAM HR Raman system (Horiba Jobin Yvon), equipped with two monochromator gratings and a charge coupled device detector (CCD). A 1,800 g/mm grating and 100 μm hole results in a 1.2 cm^{-1} spectral resolution. The 514.5 nm line of an Ar$^+$ laser is used as the ~~an~~ excitation source. Scanning electron microscopy (SEM) images are employed to evaluate the growth rate and the morphology of the surface. Adhesive tape tests (ASTM D3359) are used to qualitatively evaluate adhesion of the films to the different substrates.

3. RESULTS

3.1. Silicon Substrates

Silicon is the most commonly studied substrate for diamond hetero-epitaxial growth because of its crystal affinity. However, the importance of performing a convenient pretreatment to the substrate, in order to enhance nucleation, has been settled by several works [3-5]. The improvement of the growth rate on scratched or seeded substrates is well known [6]. In order to evaluate the importance of the most common pretreatments in the present deposition method, different surface treatments were carried out on silicon wafers. Growth rates were evaluated by SEM analysis. The performed pretreatments were as follows: *Samples A* were untreated silicon wafers cleaned with acetone in an ultrasonic bath for 30 minutes (roughness: 0.010 μm). *Samples B* were seeded with diamond crystals of 0.25 μm by an ultrasonic bath for 30 minutes (roughness: 0.019 μm). *Samples C* were manually scratched with ~~a~~ diamond paste of 0.25 μm; then they were ultrasonically cleaned with acetone several times in order to remove the diamond paste (roughness: 0.021 μm). *Samples D* were scratched with SiC; then they were ultrasonically cleaned with acetone several times in order to remove the remaining material (roughness: 0.035 μm).

Samples A, B, C, and D were processed in the deposition chamber under the same conditions for 4 hours. In all cases, diamond films were observed by optical microscopy and characterized by Raman spectroscopy.

Concerning the quality of the films evaluated by Raman and XRD, no differences were observed for different pre-treatments. The XRD spectrum result from *Sample A* is shown in Figure 3; two peaks corresponding to diamond~~s~~ (~ 44 and 75 °) can be seen along with the respective silicon substrate (~ 55 °). A low intensity peak corresponding to SiC can also be distinguished (~35 °). The Raman spectrum shown in Figure 4 presents the characteristic peak at 1,332 cm^{-1} corresponding to diamond and only a small amount of amorphous carbon is present (band around 1360-1580 cm^{-1}).

Without regard to the pretreatment methods, all the substrates were covered with continuous films. As an example, SEM micrographs corresponding to samples C and D are shown in Figure 5 for long deposition times (4 hours). However, the growth rate of the films in the first half hour of the deposition was higher on samples B, C, and D, on which scratching or seeding pretreatments were performed while only a few and isolated diamond crystals grew on the untreated silicon surface (Sample A) during the earliest stage of the process (Figure 6).

As can be seen from the presented results, this method proved to be efficient in producing good quality diamond films on silicon using a very low carbon concentration in the gas mixture and with a growth rate of around 10 μm h^{-1}.

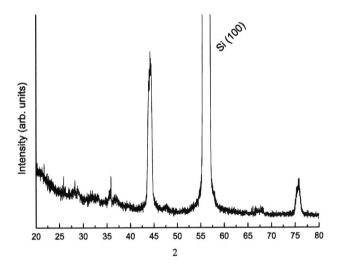

Figure 3. XRD spectrum of a diamond film deposited by on silicon.

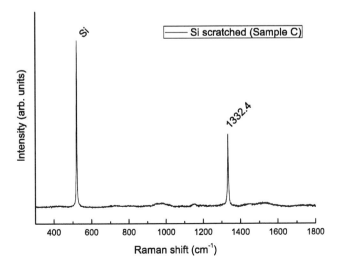

Figure 4. Raman spectrum of a diamond film deposited on a scratched silicon wafer.

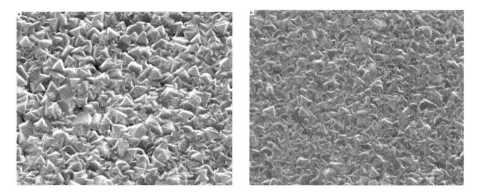

Figure 5. SEM micro-graphs of pretreated samples: diamond scratched (Sample C: left) and SiC scratched (Sample D: right).

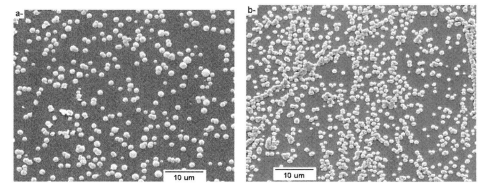

Figure 6. SEM micro-graphs of diamond crystals deposited on a Si substrate without pretreatment after 1hr deposition (Sample A: left) and on a diamond-scratched Si substrate after 10 min deposition (Sample B: right).

3.2. Stainless Steel Substrates

Diamond film deposition on steel has been one of the main research areas in diamond deposition for the last 25 years. Due to the low friction coefficient, high hardness, and exceptional wear properties, diamond appears to be an attractive material for coatings aimed to enhance the performance and durability of steel tools and devices. But direct deposition of diamond on steel substrates is, nevertheless, associated with many troubles difficult to overcome. The expansion mismatch between steel substrates and diamond films, along with the high solubility of carbon in the austenitic phase, render most CVD based diamond deposition techniques, with hydrocarbon precursors and high temperatures involved, unfeasible [2].

Graphite (instead of diamond) formation takes place at the deposition surface of steels mainly due to the low activity of carbon ($a_c<1$) in the gas phase. An excess of carbon diffusing into the metallic phase leads to the formation of cementite (Fe_3C) at the surface which acts as a diffusion barrier for carbon and promotes precipitation of graphite at the surface. Iron atoms from the decomposition of cementite diffuse through the graphite layer reaching the surface where they continue to enhance the precipitation of more graphite [7,8]. To avoid graphite formation, metal and composite films have been substantially studied as barrier coatings for diamond deposition on steels [9-18]. Favorable results have also been obtained with the use of diffusion layers, with the added benefit of strong adhesion between the steel substrate and the diffused interlayer that bears the diamond film. Quality diamond films have been successfully grown on steel by creating chromium, boron, and carbon diffusion layers [19-21].

Silicon diffusion is known, and has regularly been employed, to enhance hot corrosion resistance on stainless steels [22-24]. In the present work, continuous and adherent diamond films have been successfully deposited on AISI 304 and AISI 316 stainless steel by forming silicon diffused surface layers on the substrates before the deposition process.

Hereafter, we present a siliconization process as a surface pretreatment to successfully deposit diamond films on stainless steel substrates.

A thin layer of amorphous-silicon (a-Si) was deposited on steel substrates by Ion Beam Deposition. Silane (SiH_4) was employed as a source of silicon ions; the deposition energy was 2 keV.

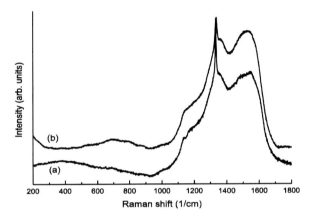

Figure 7. Raman spectra of a diamond film deposited on AISI 316 (a) and AISI 304 (b) stainless steel [25].

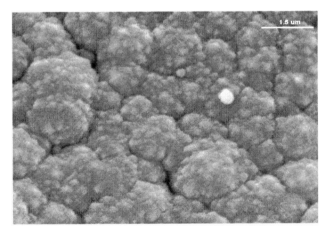

Figure 8. SEM micrograph of a film deposited on AISI 316 SS. The thickness of the film is 3 μm and the deposition rate is 1μm/h [25].

The deposited layer was about 0.1 μm thick and it was characterized by Raman spectroscopy. Then the steel was heated up in a vacuum chamber ($< 10^{-6}$ mbar) for one hour, under the same thermal conditions used for diamond deposition (~ 800 °C) with the aim of creating a diffusion layer. The aforesaid process was repeated until an undiffused silicon layer could be detected on the surface after the thermal treatment [25].

After the first diffusion treatment, a Raman analysis did not detect the presence of silicon. The process of a-Si deposit and annealing was repeated twice or three times until Raman spectra revealed a silicon layer on the steel surface. This indicated that silicon saturation had been reached in the austenitic phase close to the surface.

Diamond deposition was performed under the conditions described above at a growth rate of 1 to 3 μm h^{-1}. The Raman spectra of the films deposited on AISI 316 SS and AISI 304 SS are shown in Figure 7. As can be seen, the characteristic diamond peak at 1,332 cm^{-1} is present together with two broad bands centered at 1,350 and 1,550 cm^{-1} indicating the presence of amorphous carbon with different degrees of graphitization. A slight insinuation of bands around 1,140 and 1,490 cm^{-1} which has been attributed to acetylene C-H chains is also noticeable [26]. A SEM micrograph of a film obtained on AISI 316 SS is shown in Figure 8.

In order to verify the influence of the pretreatment, deposits were performed under the same conditions on substrates on which the siliconization process was not concluded (i. e. when after the first a-Si deposition and annealing process, no silicon signal was observed by Raman spectroscopy). The obtained films consisted mainly of glassy carbon (GC). This result concurs with those obtained for untreated steel.

It has been shown that diamond can be successfully grown on stainless steel after a siliconization process. The quality of diamond films deposited on siliconized layers is lower than that of the films deposited on crystalline silicon wafers, mainly due to the presence of amorphous carbon. This fact has also been reported for diamond films grown on diffused boride layers in AISI 316 stainless steel by Buijnsters *et al* [17]. The presence of sp^2 carbon in the deposited films shows that, although considerably attenuated, iron catalytic activity stabilizing sp^2 phases is still taking place to some extent on the growing surface.

3.3. Alumina Substrates

Diamond deposition on alumina substrates has lately become a relevant subject of research, mainly due to two significant areas of application:

i   Alumina ceramic components are employed as structural materials and in wear resistant applications. Diamond coatings can improve wear resistance and lower friction on alumina surfaces and as such , broaden the scope of application of alumina components.

ii Alumina ceramics are widely used as a substrate material in integrated circuits (ICs), despite its low thermal conductivity and large dielectric constant. An outstandingly high thermal conductivity along with a low dielectric constant makes diamond a prospective choice for coating alumina substrates to achieve higher speed and power in IC applications.

However, the thickness of the deposited diamond is severely limited by the high residual stress between the diamond film and the alumina substrate, which impairs adherence of the deposited diamond layer with increasing thickness. The quality of the diamond films has also been found to affect adherence; the presence of diamond-like carbon in the diamond films reduces the bonding between deposited films and alumina substrates [27]. In order to improve film adherence, interlayers of TiC, TiN, or TiC-TiN have been used and carbon ions have been implanted to the alumina wafers to reduce the residual stress at the interfaces [27-29].

Below we describe all the procedures followed for diamond deposition on alumina substrates.

Diamond films were deposited on three types of alumina substrates: sintered alumina wafers, flexible alumina wafers, and lightly compacted alumina powder layers.

i   Sintered alumina wafers (Coors Ceramics Company) were cut and used as substrates for diamond deposition. In order to detect the presence of impurities, metal composition was measured by EDX-SEM, yielding 91, 8.1, and 0.9 at% for Al, Si, and Ca respectively.

ii Sintered laminated alumina wafers were made in our laboratory from flexible alumina samples. An Al_2O_3 plastic forming feedstock was prepared with a partially water soluble polymer as a binder. The powder used for the present study was a fine commercial Al_2O_3 powder with 99.9 % purity. The particle size was less than 1 µm and a polymethylmethacrylate solution was used in order to form an uniform slurry at room temperature [30]. The processing method was carried out by mixing the alumina powder with the organic additive and deionized water (15 -25 % in weight), and homogenizing in a mixer. The resulting mixture consisted of a characteristic plastic mass which was then formed in thin slices (1 mm) and dried at room temperature for 24 hours. These slices were roll laminated and the film obtained was pressed to 200 bar for 5 minutes, thus obtaining a plastic form of alumina which could be cut and shaped as desired. When heated in air at a rate of 5 °C/min to 1,600 °C and held at this temperature for 5 hours, it is possible to obtain a solid alumina piece. In this process there is a loss of mass as a function of temperature due to the combustion of the binder as can be seen from the thermogravimetric analysis (Figure 9).

iii The lightly compacted alumina powder layers were made by mixing high purity 99.99% alumina powder (Taimicron, Taimei Chemicals Co. Ltd.) of 0.25 µm diameter with analytical grade absolute ethanol, manually pressing the paste with the help of a stainless steel plate against a sintered alumina wafer and baking the pressed 0.5 mm alumina paste layer at 150 °C for two hours to completely evaporate the ethanol.

Commercial sintered and sintered flexible alumina samples were either processed in the deposition chamber with no previous surface treatment or seeded using 0.25 µm diamond paste and ultrasonically rinsed in ethanol. Rugosity was measured in both cases, yielding R_a=0.55 µm for commercial alumina wafers and R_a= 0.30 µm for flexible alumina samples. After treating the surface with diamond paste, rugosity values remained the same in both cases. For lightly compacted alumina powder layers, no surface pretreatment or rugosity measurements could be performed due to the softness of the alumina powder matrix.

Good quality diamond films were grown on commercial and sintered flexible alumina substrates. At the first stages of the deposition process, nucleation and growth rates were enhanced in those samples with a surface treatment, but after a 30 minutes deposition, diamond films of about 3 µm thick could be readily obtained for all samples independently of surface treatment. A typical film can be observed in Figure 10.

On lightly compacted alumina powder substrates, however, the deposition always remained restricted to isolated clusters of diamond crystals dispersed over the surface of the samples. As the composition of these substrates was identical to the composition of sintered flexible alumina samples with no previous surface treatment, the lack of coalescence to form a continuous layer could be due to the high rugosity of the surface.

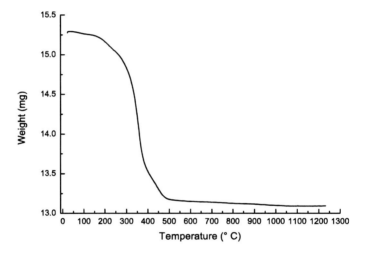

Figure 9. Thermogravimetric curve for flexible alumina.

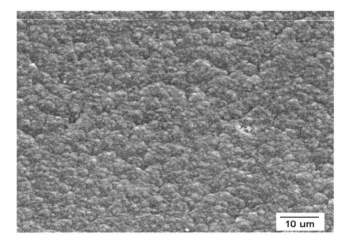

Figure 10. SEM micrograph of a diamond film on a commercial alumina substrate.

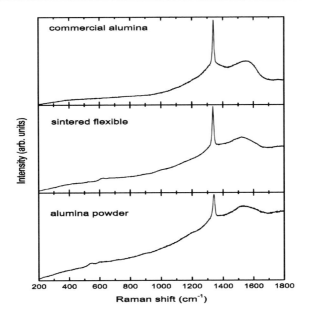

Figure 11. Raman spectra of diamond films deposited on A: commercial alumina wafers, B: sintered flexible alumina wafers, C: pressed alumina powder.

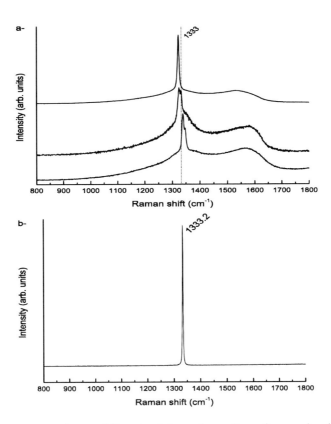

Figure 12. a- Raman spectra taken at different points on the surface of a sample with inhomogeneous structure; b- Typical Raman spectrum of any point of the same sample after an annealing process (650 °C-30 min).

Diamond quality was examined by Raman spectroscopy. The sharp peak near 1,336 cm^{-1} in the Raman spectra, together with the bands corresponding to a-C can be observed in Figure 11; no dependence on the substrate pretreatment can be noticed. After an annealing process at 650 °C in air for 30 minutes, the a-C band disappeared and the diamond peak was observed at 1,333 cm^{-1}; i.e. the peak position of stress-free natural diamond.

For samples with areas below 20 mm^2, the growth rate was particularly high (> 10 μm h^1) due to the concentration of the plasma. In those cases, a lower a-C content was present but the diamond Raman peak was observed in frequencies ranging from 1,320 to 1,340 cm^{-1}, denoting the internal stress of the films due to the inhomogeneity in the structure of the deposits.

It is well known that an isotropic compressive/tensile stress causes an up/down-shift of the Raman peak [31]. However, the presence of uniaxial stress lifts the degeneracy of the Raman active mode and a splitting can be observed [32]. This could explain the presence of frequency shifts and splitting in inhomogeneous samples, as can be seen in Figure 12a. After an annealing process at 650 °C (30 min), no a-C content is observed and the internal stress is released; the Raman spectrum becomes that of a good quality diamond film showing only one sharp peak at 1,333 cm^{-1} (Figure 12b).

Adherence of the diamond films was evaluated by means of the adhesive tape test in commercial and sintered flexible alumina samples. No rests of the deposit could be observed in the tape for 2 μm thick tested films grown on these substrates. No adherence evaluation could be performed on the lightly compacted alumina powder, due to the softness of these substrates.

3.4. Diamond Self Supporting Films

A novel method was developed to obtain self supporting diamond films on pre-sintered flexible alumina wafers. Flexible alumina samples were made as described in the previous section. The obtained plastic form is shaped and cut. Then it is heated at a rate of 5 °C/min to 250° C, when the loss of mass begins. By holding this temperature for one hour, 90% of the mass corresponding to the added polymethylmethacrylate solution was lost and a solid substrate consisting of alumina, organic compounds, and carbon that could be handled without breaking or shattering could be achieved. The pre-sintered alumina was then seeded with diamond crystals (0.25 μm) in order to enhance nucleation centers. Diamond films were deposited on these substrates under the same conditions described previously. After a deposition time of 10 hours, about 130 μm thick diamond layers were obtained. In order to easily detach the diamond film from the substrate, an annealing process was carried out. During the heating of up to 650 °C in air, the remaining carbon from the alumina was eliminated, so the substrate contracted and the diamond film could be easily detached. Finally, diamond films were carefully rinsed in 1M NaOH to eliminate any remains of Al_2O_3 powder that could be left on the bottom surface of the diamond layers. The areas of the obtained self supporting diamond layers were up to 25 mm^2.

The rugosity of the surface was 0.45 μm. Figure 13 shows SEM micrographs of a diamond self supporting film, front and side view, where thickness can be measured. A typical Raman spectrum of a self supporting diamond film is shown in Figure 14.

One of the advantages of depositing on pre-sintered flexible alumina is the possibility to obtain films with a wide range of shapes and forms. The front surface and side faces (~ 1 mm thick) of the pre-sintered alumina samples were coated with diamond; the resulting films after unmoulding can be observed from the picture in Figure 15, for two designed shapes.

Interest in self supporting CVD diamond layers has been increasing during the last decade due to the wide range of possible applications. Polycrystalline diamond films have been employed for radiotherapy and in vivo clinical radiation dosimetry because of its near tissue equivalence and chemical inertness, with the advantage of low dark currents, low sensitivity to visible light, and high carrier mobility [33-37].

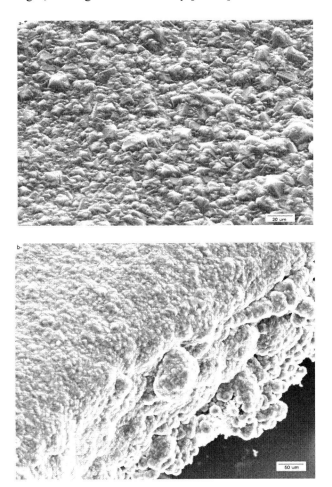

Figure 13. SEM micro-graph of a self supporting diamond film: surface (up), edge (down).

CVD diamond films have also been studied as photosensors for vacuum ultraviolet radiation [38]. Their use for spectral windows, including X-ray windows for submicron lithography of integrated circuits and X-ray spectroscopy [39], as well as in infrared synchrotron applications [40], has also been evaluated. Doped diamond self supporting layers could eventually be employed for diamond based heaters and thermometers capable of withstanding radiative or corrosive environments [41]. Furthermore, application of self supporting diamond films as parts of microelectromechanical systems (MEMS) have also been extensively studied [42,43].

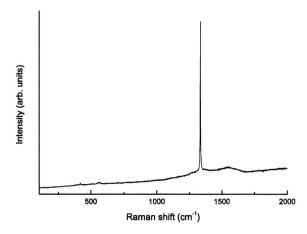

Figure 14. Raman spectrum of a self supporting diamond film.

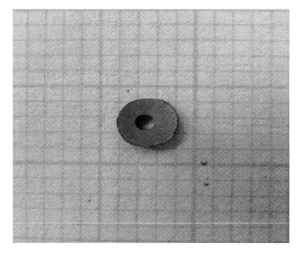

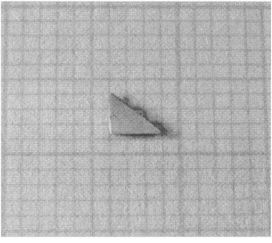

Figure 15. Pictures of self supporting diamond films.

CONCLUSION

We have presented in detail a deposition method for diamond films which yields high growth rates (up to 10 μm h^{-1}) on metallic and ceramic substrates with a low methane/hydrogen concentration (0.2%). We have shown results on silicon wafers, stainless steel, and commercial and flexible sintered alumina. It has been pointed out that surface pre-treatments favor the formation of nucleation centers and increase the growth rate. Particularly on steel, it becomes imperative to perform a previous procedure to successfully grow diamond films; a siliconization process reveals itself as a good alternative to prior presented methods. On alumina samples, stress on diamond films was evidenced by Raman spectroscopy. This effect is particularly noticeable on films with higher growth rates. An annealing process was shown to release the stress. The pre-sinterization process presented here for flexible alumina permits the design of any shape for the self supporting diamond films (thickness c.a. 100 μm). With this method 3D surfaces can be readily obtained thanks to the malleability of the alumina substrate whereas conventional self supporting films are generally grown on silicon wafers.

ACKNOWLEDGMENT

We would like to thank Mariana Rosenbusch for SEM analysis, Alicia Petragalli for X-ray measurements, and Marcelo Igarzábal, Juan Carlos Suárez Sandín, and Alejandro Zavala for their technical assistance.

REFERENCES

[1] Smith G. J. *S. Afr. Inst. Min. Metall.* 2003, 103-9, 529-534.
[2] May P.W. *Philos. Trans. R. Soc. Lond.* A 2000, 358/1766, 473-495.
[3] Flöter A.; Mainz B.; Stiegler J.; Falke U.; Schulze S.; Deutschmann S.; Schaarschmidt G. *Diamond Relat. Mater.* 1994, *3*, 1097-1102.
[4] Schelz S.; Borges C.F.M.; Martinu L.; Moisan M. *Diamond Relat. Mater.* 1997, *6*, 440-443.
[5] Wang S.G.; Qing Zhang ; Yoon S.F.; Ahn J.; Wang Q.; Yang D.J.; Huang Q.F.; Rusli; Tang W.Z.; Lu F.X. *Diamond Relat. Mater.* 2002, *11*, 1683–1689.
[6] Buijnsters J.G.; Vázquez L.; ter Meulen J.J. *Diamond Relat. Mater.* 2009, *18*, 1239–1246.
[7] Chii Ruey Lin; Cheng Tzu Kuo *Diamond Relat. Mater.* 1998, *7*, 903-907.
[8] Endler I.; Leonhardt A.; Scheibe H.-J.; Born R. *Diamond Relat. Mater.* 1996, *5*, 299-303.
[9] Buijnsters J.G.; Shankar P.; Fleischer W.; van Enckevort W.J.P.; Schermer J.J.; ter Meulen J.J. *Diamond Relat. Mater.* 2002, *11*, 536-544.
[10] Schäfer L.; Fryda M.; Stolley T.; Xiang L.; Klages C.-P. *Surf. Coat. Technol.* 1999, *116-119*, 447-451.
[11] Glozman O.; Hoffman A. *Diamond Relat. Mater.* 1997, *6*, 796-801.

[12] Bareiβ J.C.; Hackl G.; Popovska N.; Rosiwal S.M.; Singer R.F. *Surf. Coat. Technol.* 2009, *201*, 718-723.

[13] Hong-Xia Zhang; Ying-Bing Jiang; Si-ze Yang; Zhangda Lin; Ke-an Feng *Thin Solid Films* 1999, *349*, 162-164.

[14] Li Y.S.; Tang Y.; Yang Q.; Xiao C.; Hirose A. *Int. J. Refract. Met. Hard Mater.* 2009, *27*, 417-420.

[15] Bareiβ C.; Perle M.; Rosiwal S.M.; Singer R.F. *Diamond Relat. Mater.* 2006, *15*, 754-760.

[16] Gowri M.; Li H.; Schermer J.J.; van Enckevort W.J.P.; ter Meulen J.J. *Diamond Relat. Mater.* 2006, *15*, 498-501.

[17] Buijnsters J.G.; Shankar P.; Gopalakrishnan P.; van Enckevort W.J.P.; Schermer J.J.; Ramakrishnan S.S.; ter Meulen J.J. *Thin Solid Films* 2003, *426*, 85-93.

[18] Huang H.L.; Lee T.Y.; Gan D. *Mater. Sci. Eng. A* 2006, *422*, 259-265.

[19] Nishimoto A.; Nakao K; Ichii K.; Akamatsu K. in *Novel Materials Processing by Advanced Electromagnetic Energy Sources Proceedings of the International Symposium on Novel Materials Processing by Advanced Electromagnetic Energy Sources* (S. Miyake ISBN: 978-0-08-044504-5 Osaka, Japan 2005) 433-436.

[20] Grunling H.W.; Bauer R. *Thin Solid Films* 1982, *95*, 3-20.

[21] Hiroyuki Aikyo; Ken-ichi Kondo *Jpn. J. Appl. Phys.* 1989, 28, L1631-L1633.

[22] Sirenko A.A.; Fox J.R.; Akimov I.A.; Xi X.X.; Ruvimov S.; Liliental-Weber Z. *Solid State Commun.* 2000, *113*, 553-558.

[23] Swain B.P. *Surf. Coat. Technol.* 2006, *201*, 1132-1137.

[24] Perez J.M.; Villalobos J.; McNeill P.; Prasad J.; Cheek R.; Kelber J.; Estrera J.P.; Stevens P.D.; Glosser R. *Appl. Phys. Lett.* 1992, *61/5*, 563-565.

[25] Álvarez F.; Reinoso M.; Huck H.; Rosenbuch M. *Appl. Surf. Sci.* 2010, *256*, 3962-3966.

[26] Fernandes A.J.S.; Neto M.A.; Almeida F.A.; Silva R.F.; Costa F.M. *Diamond Relat. Mater.* 2007, *16*, 757-761.

[27] Fan W.D.; Wu H.; Jagannadham K.; Goral B.C. *Surf. Coat. Technol.* 1995, *72*, 78-87.

[28] Wang L-J.; Xia Y-B.; Fang Z-J.; Zhang M-L.; Shen H-J. *Chin. Phys. Lett.* 2004, *21/6*, 1161-1163.

[29] Qingfeng Su; Jianmin Liu; Linjun Wang; Weimin Shi; Yiben Xia *Diamond Relat. Mater.* 2006, 15, 1550–1554.

[30] Mesquita F.A.; Morelli M.R. *J. Mater. Process. Technol.* 2003, *143-144*, 232-236.

[31] Sherman W.F. *J. Phys. C: Solid State Phys.* 1985, *18*, L973-L978.

[32] Pandey M.; D´Cunha R.; Tyagi A.K. *J. Alloys Compound* 2002, *333*, 260-265.

[33] Bergonzo P.; Foulon F.; Marshall R.D.; Jany C.; Brambilla A.; McKeag R.D.; Jackman R.B. *Diamond Relat. Mater.* 1999, *8*, 952–955.

[34] Ramkumar S; Buttar C.M.; Conway J.; Whitehead A.J.; Sussman R.S.; Hill G.; Walker S. *Nucl. Instrum. Methods Phys. Res. Sect. A* 2001, *460/2-3*, 401-411.

[35] Buttar C.M.; Airey R.; Conway J.; Hill G.; Ramkumar S.; Scarsbrook G.; Sussmann R.S.; Walker S.; Whitehead A. *Diamond Relat. Mater.* 2000, *9/3-6*, 965-969.

[36] Conte G.; Somma F.; Niké M. in *3rd International Conference "Novel Applications of Wide Bandgap Layers"* ISBN 0780371364; Editors Jan Szmidt J., Werbowy A.; Institute of Microelectronics and Optoelectronics, Faculty of Electronics and Information Technology, Warsaw University of Techology (Warsaw). 2001, pp 177 – 178.

[37] Cirrone G.A.P.; Cuttone G.; Rafaele L.; Sabini M.G.; De Angelis C.; Onori S.; Pacilio M.; Bucciolini M.; Bruzzi M.; Sciortino S. *Nucl. Phys. B* 2003, *125*, 179-183.
[38] Uchida K.; Ishihara H.; Nippashi K.; Matsuoka M.; Hayashi K. *J. Light Vis. Environ.* 2004, *28-2*, 97-103.
[39] Xuantong Ying; Jinlong Luo; Peinan Wang; Mingqi Cui; Yidong Zhao; Gang Li; Peiping Zhu *Diamond Relat. Mater.* 2003, 12/3-7, 719-722.
[40] Sussmann R.S.; Pickles C.S.J.; Brandon J.R.; Wort C.H.J.; Coe S.E.; Wasenczuk A.; Dodge C.N.; Beale A.C.; Krehan A.J.; Dore P; Nucara A.; Calvani P. *Il Nuovo Cimento* 1998, *120D/4*, 503-525.
[41] Csorbai H.K.; Kovách G.; Fürjes P.; Csíkvári P.; Sólyom A.; Hárs G.; Kálmán E. *Mater. Sci. Forum* 2007, *537*, 145-150.
[42] Zongliang Cao; Dean Aslam *Diamond Relat. Mater.* 2010, *19*, 1263–1272.
[43] Altukhov A.A.; Afanas'ev M.S.; Kvaskov V.B.; Lyubchenko V.E.; Mityagin A. Yu.; Murav'ev E.N.; Pomortsev L.A.; Potapov V.A.; Spitsyn B.V. *Inorg. Mater.* 2004, *40/S1*, S50-S70.

Chapter 7

BORON DOPED CVD DIAMOND FILMS GROWN ON CERAMIC SUBSTRATES

Lívia Elisabeth Vasconcellos de Siqueira Brandão, Shay Reboh, Fabrício Casarin, Rafael Fernando Pires, Paulo Pureur Neto and Naira Maria Balzaretti
PGCIMAT, Universidade Federal do Rio Grande do Sul,
IF, Universidade Federal do Rio Grande do Sul
Porto Alegre, RS, Brazil

ABSTRACT

Boron doped diamond films have been grown adhered to silicon substrates by chemical vapor deposition (CVD) using boron containing gases that present high degree of toxicity. This chapter aims to investigate and discuss the boron incorporation into self-standing CVD diamond films grown on ceramic substrates. It was shown that it is possible to grow free-standing boron doped CVD diamond films on partially stabilized zirconia substrates using boron powder as the source for doping. Both surfaces of the diamond films, the one in contact with the substrate (smooth surface) and the one in contact with the plasma during deposition (rough surface), were analyzed by X-ray diffraction (XRD), Raman spectroscopy, scanning electron microscopy (SEM), transmission electron microscopy (TEM) and focused ion beam microscopy (FIB). The electric behavior of the doped films was evaluated by resistivity techniques by the 4-point probe method. For comparison, non-doped diamond films were also investigated by the same techniques. Results from Raman spectroscopy showed the Fano component, corroborating the boron incorporation into the film structure with concentration up to $\sim 10^{20}$ cm^{-3}, whereas no significant change in the lattice parameter of diamond was detected by X ray diffraction. The measurement of the resistivity as a function of temperature confirmed the semiconductor behavior, as expected for p-type diamond. For both doped and undoped films, the micrographs obtained by electron microscopy showed the presence of crystalline defects like twinning and preferential orientation. However, high resolution transmission electron microscopy revealed the presence of nanometric sized clusters, approximately 2 nm size, only for the boron-doped films, probably related to the segregation of boron powder used during the diamond film deposition process.

INTRODUCTION

Doping of the CVD Diamond Film with Boron

Polycrystalline diamond films grown by chemical vapor deposition (CVD) have been widely studied in the last years since they congregate unique properties suitable for several technological applications. Diamond is a semiconductor with a very high band gap (5,5eV) and, like silicon, it can be doped with donor and acceptor impurities such as boron[B] and nitrogen[N] [1-5]. The doping of CVD diamond films can be done *in situ* using the method of dissolution of a dopant in a solvent during the growth of the film or *ex situ* by thermal diffusion or ionic implantation after the growth of the film. In this chapter, the discussions will be restricted to *in situ* doping with boron.

The boron containing source can be solid, liquid or gaseous. The only gaseous form of boron at room temperature is diborane which is highly toxic [6-9]. Examples of solid sources used for boron doping include boron powder and boron trioxide. Liquid sources include boric acid, cyclic organic borinate ester and trimethylborate [10]. The non-gaseous boron sources must be heated or dissolved to enhance their vapor pressure.

Diamond has an extremely high atomic concentration, [C] = 1,763 x 10^{23} cm^{-3}, and, therefore, only small atoms, such as B and N, can be incorporated as substitutional impurities into the diamond lattice. Recently it was observed that, due to the incorporation of boron as substitutional impurities into CVD diamond films, they changed from insulator to metallic and even superconductor, depending on the doping level. For boron concentration in the range between 10^{17} to 10^{19} cm^{-3}, the energy level introduced by the boron impurity is located ~0.37 eV above the valence band edge [1-5, 11-13]. When the boron concentration is higher than 10^{20} cm^{-3}, it was observed an insulator-metal transition [14]. For very high boron concentration, diamond becomes superconductor [6-9].

The incorporation of substitutional boron into the diamond lattice during the deposition process depends on the orientation of the grain growth. Spytsin *et al* [15] observed that the incorporation of boron was larger in the crystallographic {111} family of diamond grains. Samlenski *et al* [16] confirmed this result by quantifying the concentration of boron in homoepitaxial films using the technique of nuclear reaction (NRA). Their results showed that the probability of boron incorporation was an order of magnitude larger for the {111} family that for the {100} family. Similar results were obtained by Locher *et al* [17] who analyzed boron doped polycrystalline diamond films of different textures by secondary ion mass spectrometry (SIMS). Umezawa *et al* [18] observed defects like boron dimers (B-B) in homoepitaxial diamond films, mostly in {111} families. They suggested that when there was a high concentration of boron, the growth of (100) planes may contain boron-carbon composites and, during the growth, they could be deposited and originate a surface of dimers that act as growth steps. The B-B pairs would be formed with equal probability along the [111] direction, expanding the lattice homogeneously. On the other hand, for (111) planes, B-B pairs would be prohibited along the [111] direction and could only be formed with equal probability along the [$\bar{1}$11], [1$\bar{1}$1] and [11$\bar{1}$] directions. Consequently, a higher density of B-B pairs would be formed on {111} families than on the {100} families [18].

According to Bourgeois [19], the energy required for interstitial defects of boron in diamond is greater than the energy required to create dimmers of boron [20]. Computer simulations showed that the B-C bonds are elongated and the C-C bonds are compressed in the presence of such defects. Based on these results, the highly boron doped diamond can be described by a random distribution of isolated substitutional boron atoms containing some dimers.

According to Werner and Locher [10], the texture of diamond film can be ruled by a single parameter, α, which determines the morphology of the diamond grain, the texture of the polycrystalline film and, also, the stability against the formation of structural defects such as twinning. Experimentally, it depends of the concentration of carbonaceous gas involved in the process, the temperature of the substrate during deposition and the impurities, such as boron. It is defined by [10]:

$$\alpha = \sqrt{3}\frac{v_{100}}{v_{111}} \qquad (1)$$

where v_{100} is the growth rate of (100) planes and v_{111} is the growth rate of (111) planes. Wild et al [21] observed that (100) planes did not contain twinning defects when 2< α < 3 while (111) planes did not contain twinning defects when 1 < α < 1.5. Several studies have been done to investigate the effect of boron incorporation into diamond structure [22-25]. For instance, it is known that boron decreases the parameter α by increasing the growth rate of (111) planes [10] and, therefore, changes the texture of the diamond film.

Boron-doped diamond films produced by CVD are usually grown strongly adhered on silicon substrates using boron containing gaseous, liquid or solid sources. The total amount of boron incorporated into CVD diamond films can be up to 10^{21} cm^{-3} without significant deterioration of the structural quality of the diamond film.

In previous works it was shown that partially stabilized zirconia (PSZ) was a very good candidate for growing free-standing non-doped CVD diamond films [26] and boron doped diamond films using solid boron source for doping [27].

In this chapter the effect of boron incorporation on diamond films grown on zirconia substrates was investigated by Raman spectroscopy, X-ray diffraction (XRD), scanning electron microscopy (SEM), transmission electron microscopy (TEM) and focused ion beam microscopy (FIB). Previous investigations about the microstructure of undoped and doped diamond films confirmed the presence of defects and preferential orientation of the diamond grains at the surface of the film that was in contact with the plasma during the growing process [28-33]. The electrical behavior of the boron-doped films was also investigated by 4-point probe method [27].

EXPERIMENTAL METHODS AND MATERIALS

Diamond Films

The boron doped diamond films were grown in a microwave assisted CVD reactor (MWCVD) ASTEX AX5400 (2.45 GHz). The gas mixture contained 99% of hydrogen and 1% of methane at 70 torr. The microwave power was 2.5 kW and the substrate temperature was in the range between 700°C and 800°C. The thickness of the films was in the range between 5 μm to 10 μm for a deposition time of 4 hours.

Lucchese *et al* [26, 34] showed that it was possible to grow high quality CVD diamond films on partially stabilized zirconia (PSZ) substrates and that the film did not remain adhered to the substrate after the deposition process. Moreover, the same substrate could be subsequently used for several times while the conventional silicon substrate has to be chemically etched after every deposition process to release the diamond film. After the first deposition on zirconia substrates, the white color of zirconia changed to metallic due to the formation of a thin layer of zirconium carbide (ZrC) on the surface of the substrate in contact with the plasma. In this work it was used PSZ containing 3 mol% of yttria covered by a thin layer of ZrC from previous depositions processes. The substrates were cylinders of 2.6 cm diameter and 2 mm thickness submitted to pre-treatments before deposition, according to Table 1. Amorphous boron powder was placed around the substrate on the molybdenum sample holder inside the CVD reactor. For comparison, a control group (Sample A) without boron was also investigated.

Analytical Techniques

X-ray diffraction (XRD) analysis was performed with Cu-K$_\alpha$ (λ = 1.5418 Å) using a graphite monochromator in a D500 Rigaku diffractometer. Raman spectroscopy with He-Ne and Ar excitation sources was used to verify the boron incorporation and the internal stress of the diamond films following the rule proposed by Grimsditch *et al* [35, 36]. The resistivity of the boron doped films was studied using the Physical Property Measurement System\Model 6000-Quantum Design – resistivity module by the 4-point probe method.

The morphology of the diamond films was investigated by scanning electron microscopy (SEM), transmission electron microscopy (TEM) and focused ion beam microscopy (FIB). Ion milling was used to prepare the samples for transmission electron microscopy. A JEOL JEM 2010 operating at 200 kV was used to obtain high resolution transmission electron microscopy images (HRTEM). SEM and FIB microscopies were performed in a Multibeam JEOL JIB-4500 system.

Table 1. Sample identification. The smooth surface of the free-standing film was in contact with the substrate during the deposition process while the rough surface of the film was in contact with the plasma

Sample ID	Diamond film feature	Pre-treatment
As	Undoped – smooth surface	Polishing with diamond paste (grain size 1 μm)
Ar	Undoped – rough surface	
Bs	Lightly doped – smooth surface	Polishing with diamond paste (grain size 1 μm) + boron powder around the substrate during deposition
Br	Lightly doped – rough surface	
Cs	Highly doped – smooth surface	Polishing with diamond paste (grain size 1 μm) and with amorphous boron powder + boron powder around the substrate during deposition
Cr	Highly doped – rough surface	

RESULTS AND DISCUSSION

Raman Spectroscopy

Raman spectra of boron doped diamond films with boron concentrations above 10^{20} B.cm^{-3} show significant changes compared to the Raman spectra of undoped films, related to the onset of metallic conductivity, a consequence of high doping. These modifications can also be seen in a moderate way in the spectra of the films that incorporate boron concentrations lower than 10^{20} B.cm^{-3}. The symmetric Lorentzian peak at ~1332 cm^{-1} characteristic of undoped diamond changes to an asymmetric Fano-like lineshape due to a quantum mechanical interference between the zone-centre Raman-active optical mode and the continuum of electronic states induced by the high concentration of dopant [37], indicating a strong signature of the existence of electron – phonon coupling. Another feature related to the boron incorporation on diamond films is the shift of the diamond peak to lower wavenumbers with increasing boron concentration, accompanied by two broad bands around 500 cm^{-1} and 1200 cm^{-1} [38-43]. The origin of these bands is not fully understood. There are both empirical and *ab initio* studies about the submission of highly doped diamond films to deuterium plasma, in which such bands are detected and correlated to the failure of chemical bonds between boron and hydrogen (B$_n$H$_n$) of boron clusters [19,44]. Other experimental studies associate the origin of these local bands to vibrational modes of boron dimers [37] or to the presence of sp^2 carbon in the lattice [46].

For very high boron concentration, these bands dominate the Raman spectrum suppressing the zone-centre phonon. According to Bernard *et al* [38], the peak at ~500 cm^{-1} can be fitted with a combination of Gaussian and Lorentzian line shapes and the Raman shift ω of the Lorentzian lineshape is approximately related to the boron content [B] in cm^{-3} by:

$$[B] = 8.44 \times 10^{30} \, e^{-0.048\,\omega} \qquad (2)$$

where ω is in cm^{-1}.

Figure 1 shows the Raman spectra for doped and undoped diamond films [27]. The measurements were performed on both surfaces of the free-standing as grown films: the smooth surface and the rough surface. The peaks at (1333.7 ± 0.1) cm^{-1} and $(1334.2 \pm 0,1)$ cm^{-1} shown in the inset of Figure 1a correspond to both surfaces of the undoped diamond film, respectively. The broad band between 1440 and 1470 cm^{-1} is related to non-diamond carbon in grain boundaries [45, 47-51]. The small difference between the spectra of both surfaces suggests the structure of the film was homogeneous along its thickness.

Figs. 1a clearly shows the abrupt changes in the Raman spectrum due to the boron incorporation, related to the Fano-like line shape and the bands at ~500 cm^{-1} and 1200 cm^{-1}, indicating the diamond film deposited on PSZ substrate polished with diamond paste and boron powder was highly doped (sample C in Figure 1a). This effect also exists but it is less pronounced on the film deposited on PSZ substrate polished only with diamond paste (sample B in Figure 1a). In both cases, boron powder was placed around the substrate in the CVD reactor. Table 2 shows the Raman peak positions and full width at half maximum (FWHM) for both surfaces of the boron doped diamond films compared to the undoped film, evidencing the shift to lower wavenumbers of the diamond peak. The fourth column of Table

2 shows the intrinsic stress of the free-standing diamond films calculated according to the rule proposed by Grimsditch *et al* [35, 36]. For all cases, the stress was negligible and, for the doped films the shift to lower wavenumbers of the diamond peak corresponded to a tensile stress [52]. The fifth column shows the boron concentration calculated using Eq. 2.

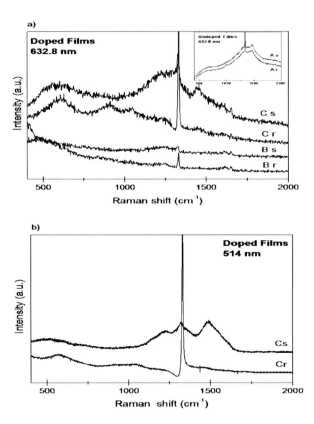

Figure 1. (a) Raman spectra measured with HeNe excitation laser (632.8 nm). The inset shows the spectrum of the undoped diamond film. The Raman spectrum of the lightly doped (sample B) and highly doped (sample C) films are shown for both smooth (s) and rough (r) surfaces. (b) Raman spectrum of both surfaces of the highly doped diamond film (C) measured with Ar excitation laser (514 nm). [27]

Table 2. Data obtained from the Raman spectra of collection of samples A, B and C [27]

Sample ID	Raman Peaks(cm^{-1})	FWHM (cm^{-1})	Intrinsic Stress (GPa)	[B] (cm^{-3}) by Eq. 2
As	1333,7±0,1	7 ± 0,03	0,4	0
Ar	1334,2±0,1	7,7 ± 0,03	0,6	0
Bs	1326,5±0,8	13,7±0,25	-1,8	--
Br	1331,7±0,1	8,6 ± 0,05	-0,2	--
Cs (HeNe)	1328,8±0,9	9,67±0,59	-1,1	~ 10^{19}
Cr (HeNe)	1331,7±0,2	7,44 ± 0,4	-0,2	~ 10^{18}
Cs (Ar)	--	--	--	~ 10^{19}
Cr (Ar)	1330,2±0,2	8,1 ± 0,4	-0,7	~ 10^{18}

In contrast with the undoped film, the Raman spectra shown in Figure 1a indicated the doped films were not homogeneous along their thicknesses. Moreover, according to Table 2, the boron concentration on the smooth surfaces seems to be higher than on the rough surface as should be expected due to the presence of boron at the surface of the substrate during deposition. The intensity of the band at ~500 cm^{-1} for sample B was too low to enable the calculation of the boron concentration using Eq. 2.

Bourgeios [19] reported that Raman spectra excited by ultraviolet radiation did not exhibit the Fano line profile. Other researchers have also observed that the appearance of Fano line depends on the wavelength of the excitation laser [17, 41, 53-56]. Figure 1b shows that the Raman spectra for both surfaces of the highly doped diamond film measured with an Ar laser (514 nm) is very similar to the spectra measured with a HeNe laser (632.8 nm), showing, also, the Fano line.

Structural Characterization

Figure 2 shows the X ray diffraction patterns for both surfaces of the undoped and doped diamond films. Bourgeois [19] investigated homoepitaxial boron doped diamond films and noted an increase in the lattice parameter of the {100}, {110} and {111} families of diamond. The results shown in Figure 2 correspond to (111), (220) and (311) diffraction planes and it was not observed any significant change in the positions and FWHM of the diffraction peaks related to the incorporation of boron in the diamond structure. The incorporation of boron was confirmed by the results from Raman spectroscopy.

Figure 3a shows the FIB image of the zirconia substrate obtained by channeling of the gallium ions due to the different crystallographic orientations of the zirconia grains. Figures 3b and 3c show the SEM and FIB images, respectively, of the surface of the doped diamond film that was in contact with the substrate. It is possible to observe that this surface of the film replicates the topography of the zirconia substrate, including grain boundaries and straight lines induced by the mechanical polishing.

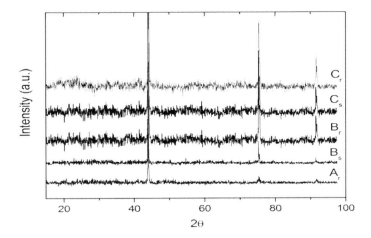

Figure 2. X-ray diffraction patterns of undoped (A), lightly doped (B) and highly doped (C) free-standing diamond film. The subscripts s and r correspond to the smooth and rough surfaces of the film, respectively [27].

Figures 4a and 4b show FIB images of the surfaces of the undoped and boron doped diamond films that were in contact with the plasma during deposition. As can be seen, the growing faces of doped diamond grains are very smooth while the faces of the undoped diamond grains are uneven. The substrate temperature and the concentration of methane used in this work corresponded to a grow parameter α in the range between 2.5 and 3 and the images shown in Figs. 4a and 4b, both for undoped and doped films, are in agreement with the previous studies [21] in which (100) faces were stable towards twinning when 2< α < 3. Figure 4c shows high resolution details of the grain surface of the undoped diamond film.

Figure 3. a) Image of the surface of PSZ substrate obtained by FIB. b) and c) FIB and SEM images of the smooth surface of the boron doped film (Samples C), respectively.

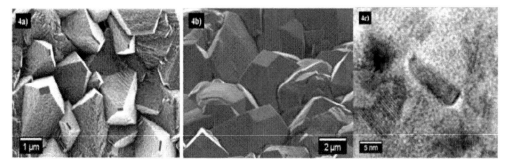

Figure 4. a) and b) FIB images of the rough surface of the undoped and doped diamond films (Samples A and C), respectively c) HRTEM image of undoped diamond film.

Figure 5 shows the stacking faults, dislocations and twinning observed by transmission electron microscopy for the undoped and doped diamond films.

Figure 6a shows an image obtained from FIB after the milling induced by gallium ions at the surface of the boron doped diamond film that was in contact with the zirconia substrate during the CVD process. As can be seen, the intragranular region was very homogeneous. Figure 6b shows an image obtained by high resolution transmission electron microscopy where it is possible to find nanometric sized clusters at the surface of the boron doped grains. These clusters were not observed in the undoped diamond grains grown at the same conditions and over the same substrate. This result indicates that these clusters may be related to the segregation of boron powder inside the diamond grains during the CVD process at the surface of the substrate.

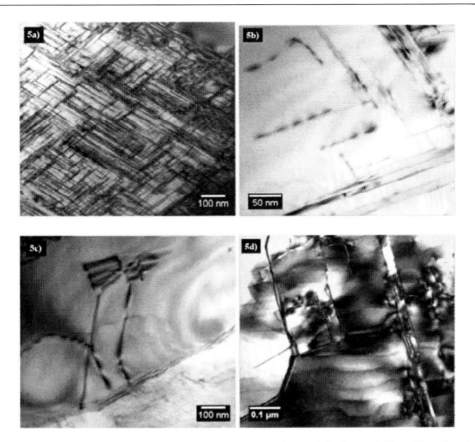

Figure 5. TEM micrographs of undoped (a and b) and doped (c and d) diamond films (Samples A and C).

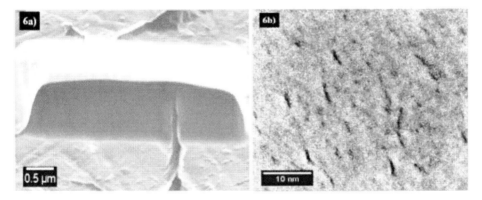

Figure 6. a) FIB Image of the intragranular region of the highly doped film (Sample C) after ion milling. b) HRTEM micrograph of nano-segregation detected in highly doped films (Sample C).

It was observed that the diamond films, both doped and undoped, were very stable against the gallium ion beam during the FIB analyses. It was very difficult to detect effects of redeposition or amorphization during the ion beam imaging.

Electrical Characterization

Figure 7 shows the resistivity of both surfaces of sample B as a function of temperature [27]. The resistivity of the undoped diamond film was higher than 10^3 Ω.m in the high temperature range. The observed decrease of the resistivity for the boron doped film with increasing temperature was expected for semiconductor materials. As can be seen, the resistivity of the rough surface of the film was one order of magnitude smaller than the resistivity of the smooth surface of the film. A possible explanation for this difference may be related to the typical grain sizes for both surfaces of the diamond film, as showed in Figs. 3 and 4. In the smooth face, the grain size was much smaller and, therefore, the density of grain boundaries was larger compared to the rough size. The resistivity measurement depends on the boron doping level in the diamond grains but, also, on the non-diamond impurities at the grain boundaries. The contribution of the grain boundaries is more relevant in the smooth surface and, as a consequence, the resistivity of this surface is higher than the resistivity of the rough size.

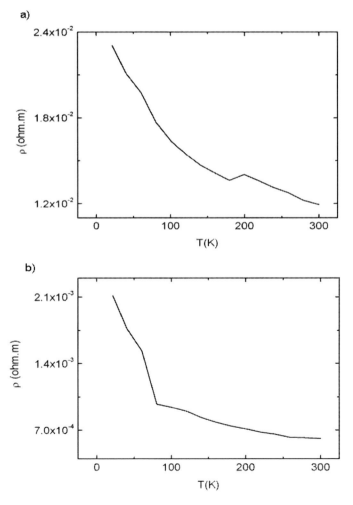

Figure 7. Temperature behavior of the resistivity for both surfaces of the slightly doped free-standing diamond film: (a) smooth surface (Bs) and (b) rough surface (Br) [27].

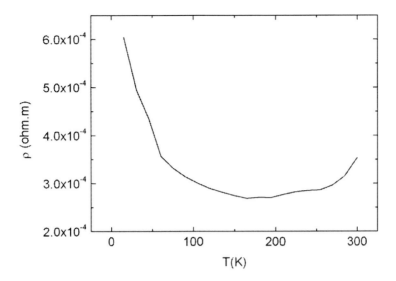

Figure 8. Temperature behavior of the resistivity of the smooth side of the highly doped free-standing diamond film (Cs) [27].

Figure 8 shows the resistivity of the smooth surface of the highly doped diamond film (sample C) as a function of temperature [27]. As can be seen, the resistivity was two orders of magnitude smaller than the smooth surface of the slightly doped diamond film (Figure 7a). Since the grain sizes of the smooth surface of both films were alike, the decrease of the resistivity should be related to higher boron incorporation inside the diamond grains in Sample C. Figure 8 also shows that the resistivity of the highly doped film started to increase for temperatures higher than 250 K, as should be expected for metallic materials, probably due to the high level of doping.

Table 3 lists the numbers of carriers obtained by Hall effect and the order of magnitude of the electrical resistivity of these films [27]. Comparing the results shown in Tables 2 and 3, the values obtained for the boron concentration from Raman spectroscopy and electrical measurements were in good agreement.

Table 3. Data obtained by Hall Effect and electrical resistivity measurements for samples A, B and C

Sample ID	Concentration of carriers (cm^{-3})	Resistivity T = 300K (Ω.m)
A	~ 9,1 x 10^{10}	3,48 x 10^{3}
B	~ 1,1 x 10^{18}	1,32 x 10^{-3}
C	~1,1 x 10^{20}	3,5 x 10^{-4}

CONCLUSION

It was possible to grow free-standing diamond films doped with boron on partially stabilized zirconia substrates. If boron powder was placed around the substrate during the CVD process and the substrate was polished with boron powder prior deposition, the free-standing diamond film was highly doped (~10^{19} - 10^{20} cm^{-3}). Raman spectroscopy was used to confirm the boron incorporation in both surfaces of the doped films. The X- ray diffraction patterns indicate the effect of boron incorporation was negligible in the diamond structure.

The diamond film surface in contact with the substrate (smooth surface) during the CVD process replicates the morphology of the zirconia substrate. The facets of the grains of the boron doped diamond films were very smooth. HRTEM images indicate the formation of nanometric sized clusters inside the boron doped diamond grains. These clusters may be related to the segregation of boron atoms that were in excess at the surface of the zirconia substrate due to the polishing prior the deposition process. Besides that the grain sizes of the smooth surface were much smaller than the grain sizes of the rough surface. This heterogeneous microstructure along the small thickness of the film was reflected on the resistivity measurements. The resistivity measured on the smooth surface of the doped films was one order of magnitude larger than the values measured for the rough surface, probably due to the higher density of grain boundaries on the smooth surface.

In conclusion, zirconia substrates combined with boron powder are suitable for growing free-standing and highly doped diamond films.

ACKNOWLEDGMENT

The authors would like to thank the financial support from the Brazilian Agency CNPq and the Electronic Microscopy Center, the Nanometric Conformation Laboratory and the Resistivity Laboratory from UFRGS for the SEM, TEM, FIB and electrical analysis.

REFERENCES

[1] Y. V. Pleskov, A. Y. Sakharova, M. D. Krotova, L. L. Bouilov, B. V. Spitsyn, *J. Electroanal. Chem.* 228 (1987) 19-27.
[2] R. Ramesham, R. F. Askew, M. F. Rose, B. H. Loo, *J. Electrochem. Soc.* 140 (1993) 3018-3020.
[3] G.M. Swain, *Adv. Mater.* 6 (1994) 388-392.
[4] C. Reuben, E. Galun, H. Cohen, R. Tenne, R. Kalish, Y. Muraki, K. Hashimoto, A. Fujishima, J. M. Butler, C. Lévy-Clément, *J. Electroanal. Chem.* 396 (1995) 233-239.
[5] J. Z. Zhu, S. Z. Yang, P.L Zhu, X. K. Zhang, G. X. Zhang, C. F. Xu, H. Z. Fan, Fresenius *J. Anal. Chem.* 352 (1995) 389-392.
[6] E. A. Ekimov, V. A. Sidorov, E. D. Bauer, N. N. Mel'nik, N.J.Curro, J. D. Thompson, S.M. Stishov, *Nature* 428 (2004) 542.
[7] Y. Takano, M. Nagao, I. Sakaguchi, M. Tachiki, T. Hatano, K. Kobayashi, H. Umezawa, H. Kawarada, *Appl. Phys. Lett.* 85 (2004) 2851-2853.
[8] T. Yokoya, T. Nakamura, T. Matsushita, T. Muro, Y. Takana, M. Nagao, T. Takenouchi, H. Kawarada, T. Oguchi, *Nature* 438 (2005) 647-650.
[9] E. Bustarret, J. Kacmarcik, C. Marcenat, E. Gheeraert, C. Cytermann, J. Marcus, T. Klein, *Phys. Rev. Lett.* 93 (2004) 237005.
[10] M. Werner, R. Locher, *Rep. Progr. Phys.* 61 (1998) 1665.

[11] J. Mort, D. Kuhman, M. Machonkin, M. Morgan, F. Jansen, K. Okumura, Y. M. LeGrice, R. J. Nemanich, *Appl. Phys. Lett.* 55 (1989) 1121-1123.
[12] G. M. Swain, R. Ramesham, *Anal. Chem.* 65 (1993) 345-351.
[13] R. M. Chrenko, *Phys. Rev.* B. 7 (1973) 4560-4567.
[14] R. Long, Y. Dai, M. Guo, L. Yu, B. Huang, R. Zhang, W. Zhang, *Diamond and Relat. Mater.* 17 (2008) 234-239.
[15] B. V. Spytsin, L. L. Bouilov, B. V. Derjaguin, *J. Cryst. Growth* 52 (1981) 219.
[16] R. Samlenski, C. Haug, R. Brenn, C. Wild, R. Locher, P. Koidl, *Diam. Relat. Mater.* 5 (1996) 947.
[17] R. Locher, J. Wagner, F. Fuchs, M. Maier, P. Gonon, P. Koidl, *Diam. Relat. Mater.* 4 (1995) 678.
[18] H. Umezawa, T. Takenouchi, Y. Takano, K. Kobayashi, M. Nagao, I. Sakaguchi, M. Tachiki, T. Hatano, G. Zhong, M. Tachiki, H. Kawarada, http://www.arXiv.org/condmat/0503303.
[19] E. Bourgeois, In Couplage électron-phonon dans les semi-conducteurs dopes et ses applications à la supraconductivité; Lyon (2008) Thesis (Physics PhD), Université Claude – Bernard – Lyon I.
[20] J. P. Goss, P. R. Briddon, *Physical Review* B 73 (2006) 085204.
[21] C. Wild, R. Kohl, N. Herres, W. Muller-Sebert, P. Koidl, *Diam. Relat. Mater.* 3 (1994) 373.
[22] A. J. Eccles, T. A. Steele, A. Afzal, C. A. Rego, W. Ahmed, P. W. May, S. M. Leeds, *Thin Solid Films* 627 (1999) 343-344.
[23] T. Shirasaki, A. Derré, M. Ménétrier, A. Tressaud, S. Flandrois, Carbon 38 (2000) 1461.
[24] O. A. Voronov, A. V. Rakhmanina, *Inorganic Materials* 29 (1993) 707.
[25] M. T. Rooney, *J. Cryst. Growth* 116 (1992) 15.
[26] M. M. Lucchese, C. L. Fritzen, A. S. Pereira, J. A. H. da Jornada, N. M. Balzaretti, *Diam. and Relat. Mater.* 14 (2005) 1605-1610.
[27] L. E. V. de S. Brandão, R. F. Pires, N. M. Balzaretti, *Vibrational Spectroscopy* 54 (2010) 84 – 88.
[28] J. Wong, E. F. Koch, C. I. Hejna, M. F. Garbauskas, *Journal of Applied Physics* 58 (1985) 3388-3393.
[29] T. Evans, C. Phaal, *Proceedings of the Royal Society* A270 (1962) 538-552.
[30] L. A. Bursill, J. L. Hutchison, A. R. Lang, N. Sumida, *Nature* 292 (1981) 518-520.
[31] P. B. Hirsch, J. L. Hutchison, J. Titchmarsh, *Philosophical Magazine Letters* A54 (1986) L49 – L54.
[32] J. C. Walmsley, A. R. Lang, M. L. T. Rooney, C. M. Welbourn, *Philosofical Magazine Letters* 55 (1987) 209-213.
[33] R. J. Zhang, S. T. Lee, Y. W. Lam, *Diamond and Related Materials* 5 (1996) 1288 – 1294.
[34] M. M. Lucchese, In Estudo Exploratório da Deposição de Filmes de Diamante em Alguns Substratos Cerâmicos; Porto Alegre (2002) Dissertation – (Physics Master's Degreee), Universidade Federal do Rio Grande do Sul.
[35] L. Bergman, R. J. Nemanich, *J. Appl. Phys.* 78 (1995) 6709-6719.
[36] M. H. Grimsditch, E. Anastassakis, M. Cardona, *Phys. Rev. B* 18 (1978) 901-904.

[37] P. W. May, W. J. Ludlow, M. Hannaway, P. J. Heard, J. A. Smith, K. N. Rosser, *Chem. Phys. Lett.* 446 (2007) 103-108.
[38] M. Bernard, C. Baron, A. Deneuville, *Diam. and Relat. Mater.* 13 (2004) 896-899.
[39] J. P.Goss, P. R. Briddon, R. Jones, Z. Teukam, D. Ballutaud, F. Jomard, J.Chevallier, M. Bernard, A. Deneuville, *Phys. Rev. B* 68 (2003) 235209.
[40] E. Bourgeois, E. Bustarret, P. Achatz, F. Omnes, X. Blase, *Phys. Rev. B* 74 (2006) 094509.
[41] P. Gonon, E. Gheeraert, A. Deneuville, F. Fontaine, L. Abello, G. Lucazeau, *J. Appl. Phys.* 78 (1995) 7059-7062.
[42] K. Ushizawa, K. Watanabe, T. Ando, I. Sakaguchi, M. Nishitani-Gamo Y. Sato, H. Kanda, *Diam. and Relat. Mater.*7 (1998) 1719-1722.
[43] A. Deneuville, C. Baron, S. Ghodbane, C.Agnès, *Diam. and Relat. Mater.*16 (2007) 915- 920.
[44] B. Lux, R. Haubner, *VDI-Berichte* 762 (1989) 61.
[45] R. J. Nemanich, J. T. Glass, G. Lucovsky, R. E. Shroder, *J. Vac. Sci. Technol. A*, 6 (1988) 1783-1787.
[46] S. Prawer, K. W. Nugent, D. N. Jamieson, J. O. Orwa, L. A. Bursill, J. L. Peng, *Chem.Phys. Lett.* 332 (2000) 93.
[47] B. Marcus, L. Fayette, M. Mermoux, L. Abello, G. Lucazeau, *J. Appl. Phys.* 76 (1994) 3463-3470.
[48] A. C. Ferrari, *Diam. and Relat. Mater.* 11 (2002) 1053-1061.
[49] M. Yoshikawa, G. Katagiri, H. Ishida, A. Ishitani, M. Ono, K. Matsumura, *Appl. Phys. Lett.* 55 (1989) 2608-2610.
[50] S. M. Huang, Z. Sun, Y. F. Lu, M. H. Hong, *Surf. Coat. Technol.* 151 (2002) 263-267.
[51] I. Pócsik, M. Hundhausen, M. Koós, L. Ley, *J. Non-Crystal. Solids*, 227 (1998) 1083-1086.
[52] H. Umezawa, T. Takenouchi, K. Kobayashi, Y. Takano, M. Nagao, M. Tachiki, T. Hatano, H. Kawarada, New Diam. *Frontier Carbon Technol.* 17 (2007) 1-9.
[53] E. Pruvost, E. Bustarret, A. Deneuville, *Diam. Relat. Mater.* 9 (2000) 295.
[54] E. Gheeraert, P. Gonon, A. Deneuville, L. Abello, G. Lucazeau, *Diam. Relat. Mater.* 2 (1993) 742.
[55] J. W. Ager III, W. Walukiewicz, M. McCluskey, M. A. Plano, M. I. Landstrass, *Appl.Phys. Lett.* 66 (1995) 616.
[56] Y. G. Wang, S. P. Lau, B. K. Tay, X. H. Zhang, *Appl. Phys.* 92 (2002) 7253.

In: Diamonds: Properties, Synthesis and Applications ISBN: 978-1-61470-591-8
Editors: T. Eisenberg and E. Schreiner, pp. 145-153 © 2012 Nova Science Publishers, Inc.

Chapter 8

SPECIFIC DEFECTS INDUCED BY MOLECULAR BEAM IMPLANTATION INTO A DIAMOND

S. T. Nakagawa[*]

Graduate School of Science, Okayama Univ. of Science,
1-1 R-dai-cho, Kitaku, Okayama 700-0005, Japan

ABSTRACT

Carbon is a leading element for innovative semiconductor devices. Here we consider one example of this; a specific defect the NV-N (nitrogen-vacancy-nitrogen) center in diamond. This is useful for the NOT logic operation for emerging quantum computers. An essential condition for these NV-N centers is that both N atoms are on a substitutional site and the intrapair distance (R_{N-N}) between them is properly separated at about 2 nm. Using a N_2 molecule as a projectile, we have examined what happens in a pure diamond after implantation by making use of a classical molecular dynamics (MD) and a crystallographic analysis tool called Pixel Mapping (PM). We found that sub-keV per atom energies are sufficient to dissociate the N_2 to an appropriate separation distance, if hot-implantation at a temperature of substrate of 900 ~ 1000 K is used.

A phase transition from a crystalline to amorphous (CA) state occurred in a few ps after ion implantation. Then the damaged crystallinity starts to recover in one ns, presumably by global phonons.

Keywords: diamond, nitrogen-doped defects, molecular dynamics simulation, molecular beam implantation, annealing procedure

---

[*] Corresponding author; S. T. Nakagawa: 1-1 Ridai-cho Okayama 700-0005, Graduate school of Science, Tel/Fax +81 86 256 9458. E-mail; stnak@dap.ous.ac.jp

1. INTRODUCTION

1.1. Carbon Allotropes; Differences from Silicon

Because of its great potential, carbon (C) is an Emerging Research Material [1]. The extraordinary properties of carbon originate from the characteristics in the atomic bonding due to outer-shell electrons. Carbon atoms ($Z = 6$) have notable differences in the bonding features compared with other heavier IV elements ($Z > 6$), as shown in Table 1. Typical crystals are shown with their space groups. The two types of bonding in carbon are two-dimensional sp^2-*bonding* and three-dimensional sp^3-*bonding*. However, sp^2-*bonding* does not always provide a stable structure for other heavier IV elements. The sp^2-*bonding due to the presence of π- electrons* forms a laminar, or bent sheet structures as graphite, fullerenes and carbon nanotubes. sp^3-*bonding* on the hand can construct a crystal called the Zincblende structure, which has rotational and translational, but not inversion symmetry as is often seen for IV elements, e.g., C (cubic diamond) and Si (crystalline Si). The same structure is also found in chemical compounds, e.g., 3C-SiC (IV elements) and GaAs (III - IV elements). Hereafter, we will refer to the cubic diamond as "*diamond*". When a sheet of graphene is folded, the dangling bonds produce a new atomic structure that causes the phase transition from semiconductor to metal transition at the folding edge. This behavior is unique to carbon and not observed in other IV elements.

The resemblance in the configuration of outer-shell electrons (ns^2np^2), where n is the principle quantum number, implies a similarity in physicochemical nature. As long as atoms have the same configuration, the interatomic potential (IP) between those atoms can have the same mathematical form but with a different set of potential parameters [2]. The IP is an indispensable tool in an empirical atomistic simulation using molecular dynamics (MD) [3]. For example, the Tersoff-type IP for a covalent material is often used for C, Si, Ge and their compounds with different sets of potential parameters.

Table 1. A comparison between C <u>atom</u> and heavier IV elements with respect to the state of *atomic bonding*. The □-graphite is a laminar structure made of completely flat atomic planes with the stacking fault of staggered structure of {AB}$_n$

<u>Unit</u>	<u>F</u>rame	$Z = 6$ (Carbon)	$Z > 6$	<u>N</u>otes
Atom	<u>C</u>onfiguration	$n = 2$	$n > 2$	Outer-shell electrons (ns^2np^2)
Molecule	*bonding*	σ-*bonding* +π-*bonding*	σ-*bonding*	σ–*bonding; sp³ bonding Forming a tetrahedral unit* π-*bonding; sp² bonding Forming a hexagonal unit*
(Cluster)	Magic numbers	C$_{60}$(cage of I_h) C$_{70}$, C$_{74}$, C$_{84}$…	×	C$_{60}$; (20 Hexagons (C$_6$) + 12 Pentagons (C$_5$))
Crystal	$(sp^2)_n$	α-graphite	×	# 194 (space group)
	$(sp^3)_n$	Cubic diamond	c-Si, c-Ge, c-GaAs	# 216 (space group), Zincblende (also # 122 (Chalcopyrite))
(Crystal defect)	Photon emission	NV center	×	Quantum entanglement at 300 K

Specific consideration is necessary for IPsfor the laminar graphite structure based on the sp^2 configuration (e.g. Brenner potential [4]), where additional terms should be introduced in the IP to describe the van der Walls force between layers, for example.

1.2. Impurities and Doped Centers in a Diamond

In natural diamond, many kinds of impurities and derivative defect centers have been reported [5]. One important impurity are nitrogen (N) atoms [6]. Depending on the concentration of N, diamond is grouped into two types, higher concentration (type-I) (> 1ppm N) or lower one (type-II) (< 1 ppm N). In the type-I, type-Ia and type-Ib are distinguished by whether nitrogen atoms are coalesced (Ia) or isolated (Ib). In the type-II, type-IIa and type-IIb are distinguished by if boron atom is present (IIb) or not (IIa). The latter is generally called pure diamond.

Some defects related to artificial dopants are used for specific functions as was first carried out by Shockley using ion implantation. A key defect in development of carbon technology is the nitrogen-vacancy (NV) center in a diamond [7], which emits a single-photon (638 nm at room temperature (RT) [8]). The N atom resides at a lattice site adjacent to a vacancy. The NV center can produce a long quantum entanglement if it has a negative charge and interacts with a nuclear spin nearby [9]. This makes it a candidate for fabricating a quantum bit (qubit). In the architecture of the quantum computer (QC), the effective nuclear spin is supplied by artificially introduced isotope of ^{13}C or ^{15}N [10].

Furthermore, in order to set up the logical circuit for a QC, a quantum "NOT gate (logic element for a QC)" is necessary. A model for the NOT has been proposed, by adding one substitutional N atom nearby the NV center, which is then called a "*NV-N*" center [11]. The "proper" intrapair distance between the two substitutional N atoms, R_{N-N} is essential, which has to be < 2.3 ~ 2.6 nm [12] or under 2.2 nm [11]. For the small number (10-50) qubits carbon technology based on NV-N centers is a frontline approach.

1.3. Methodology; Capability of Computer Simulations

In today's materials research and development (R&D), experimental observation and calculation are complementary methodologies in the scientific method. The role of material research is to select an optimal condition to obtain new function(s) by investigating the intermediate processes. Computer simulation has been an indispensable guiding-tool for R&D, prior to establishment of a new technology. In fact, a reliable simulator has been successfully developed for the silicon technology based on monovalent and atomic beams.

An event seamlessly develops in nature from the beginning to the end. Nevertheless, in principle, every observation and calculation looks at a snapshot of nature in a restricted realm, (energy, space, and time) each realm we call class. In every class an appropriate methodology exists [13]. An advantage of calculation to observation is that it can resolve a synergetic event in a complex-system, such as those caused by the irreversible correlation of elementary processes, which can overcome the limitations of a *Reductionism model*.

For the implantation doping technique, an optimal condition for a control can be the set of four key parameters [14]; they are ion species of the dopant (X) including its charge and number of constituent atoms, impact energy (E_0), fluence (Φ) (and its time differential), and substrate (or heat-bath) temperature (T_{HB}). These key factors are in principle significant throughout all the classes, although the relative importance of them depends on what class we are most interested in.

1.4. Defect Design

In doping design, not only doping itself but also annealing is necessary. The heat supply acts to retrieve the original crystallinity of the host material as much as possible, and promote the dopant's activation by incorporating it on a lattice site. MD can give us the exact position of every atom at each instant, however further crystallographic analysis is necessary to identify the presence of a specific crystalline defect [15] such as the NV-N center.

In the case of diamond annealing, the diamond phase is probable as a quasi-equilibrium state. Previously Gaebel et al. [11] produced the NV-N center in a IIa-type diamond with the intrapair distance (R_{N-N}) ~ 1.5 nm, using a N_2 beam. This result implied the original diamond structure was almost retrieved by post-annealing at 1073 K, after the bombardment with a energy of $6 \leq E_0$ (keV/atom) ≤ 14 at RT. They tuned the lowest energy corresponding to the longest R_{N-N}, by using a Monte-Carlo simulation.

Making use of a classical MD calculation combined with a crystallographic analysis called the Pixel Mapping (PM) method [16-18], we have searched for an optimal condition for (X, E_0, Φ, T_{HB}) to form the NV-N center in a diamond using much lower energy (E_0 (eV/atom)$\leqq 500$) to suppress the radiation damage. A molecular N_2 beam was implanted into a pure diamond or IIa [14, 19], and an atomic N beam into a Ib-type diamond [20]. The temperature of the heat-bath was below 1000 K, which is a little lower than the reported annealing temperature (~ 1073 K) [11]. At a very low temperature (T_{HB} = 10 K), when using N beam implanted into a Ib-type diamond, no NV-N center was found [20]. In this case the MD box included one NV center at its center: This meant athermal collisions were insufficient to produce the NV-N center. An optimal condition to produce the NV-N center was observed under the hot-implantation of sub-keV N_2 beam, with conditions of $900 < T_{HB}$ (K) $\leqq 1000$ and $100 < E_0$(eV/atom) < 500. The intrapair distance R_{N-N} was below 2 nm. The apparent discrepancy in the effectual E_0 from a previous experiment is explained separately [19].

2. METHOD

2.1. MD Simulation (Observation of Atomic Collisions in Matter)

The MD simulation solves simultaneous differential equations of motion for N atoms in an ensemble packed in a MD simulation volume V. We use the NVT frame that keeps the frame parameters of (N, V, T_{HB}) invariant. The necessary conditions to solve the N equations are as follows: 1) N should be sufficiently large so that the volume V still leaves a margin for the collision cascade to be developed by a projectile. The box size is virtually extended outward by mirroring using the periodic boundary condition. This idea of a super lattice is analogous to the reduced Brillion zone in the *k*-space for electrons in a crystal. 2) The time difference (Δt) to be used in a difference equation should be nontrivially small, to minimize the numerical integration error. 3) The external-force term in each equation is taken to be the derivative of the potential field at each atom, which is made up from the superposition of IPs from all other atoms where the IP is non-zero. 4) A boundary condition is applied to atom's velocity by "*velocity (v) scaling*" so that the value of the bulk temperature *T* is regulated so as to approach closer to T_{HB}. The bulk temperature is determined by $T = (2/3/k_B) \Sigma_i (m_i v_i^2)/2/ N$, where k_B the Boltzmann factor, and the summation goes over those *N* atoms interacting with the wall of the heat-bath.

The above mathematical conditions were included in the MD simulation [21]. These conditions corresponded to; $100 < E_0$ (eV/atom)≤ 500, *X* is the molecular N_2 beam implanted into IIa-type diamond, Φ is a single ion for the MD box. The heat-bath temperature is 300 $\leq T_{HB}$ (K) $\leq$ 1000. The largest volume of a MD cubic box is $V(nm^3) = 9.8 \times 9.8 \times 9.8$, containing $N = 46656$ C atoms. Before starting the impact simulation, the system was brought into equilibrium with the heat-bath by annealing for 3 ps ("$T \rightarrow T_{HB}$"). This was necessary because the IP to be used in a MD ignores the thermal vibration in a solid at $T=T_{HB}$. Strictly speaking, the longest time of MD simulation (~ ns) is still too short to examine the complete annealing process (> ms) [13, 23, 24]. Nevertheless, we can see the significant change in the early stage of relaxation that goes toward a quasi-equilibrium state.

2.2. PM Analysis (Observation of Crystalline State of Matter)

We have developed the PM method [16] to identify crystalline defects [17] such as point defects, linear defects with crystallographic orientations, planar defects with Miller indexes, and amorphization using the long-range order (LRO) parameter [16-18]. The PM is available for a ternary ($A_l B_m C_n$) crystal having cubic symmetry. The complicated atomic locations in these crystals implies the LRO parameter is composed of many components (8~16) [18]. However, in this case where the CA transition is solely studied, these components could be lumped into one single *global LRO parameter* [19].

The global LRO parameter depicts the degree of crystallinity over the entire space of the material concerned [16, 17]. We also extracted the depth dependence of this parameter, $LRO(n_z)$ [19]. The "n_z" is the number of a slab counted from the surface when the MD box was sliced parallel to the surface and perpendicular to the *z*-axis. Here we take its thickness of 0.5 nm. Our largest MD box has a volume of 9.8 × 9.8 × 9.8 nm. Therefore, *LRO* and $LRO(n_z)$ parameters represent the averages of 4.7×10^4 and 2.4×10^3 atoms, respectively.

3. RESULTS

3.1. Multiple Collisions Caused by a Low-Energy Molecular Beam

We had found the weak dependence of implantation energy on the intrapair distance of "$<R_{N-N}>$" when a N_2 beam was implanted into a pure diamond C(001) at RT ($T_{HB} = 300$ K) with the partial energy $E_0 = 100, 200, 300, 500$ (eV/atom) [14]. As E_0 increased from 100 to

500 eV/atom, the corresponding $<R_{N-N}>$ gradually increased from 1.76 to 2.45 nm. This result implies that the sub-keV energies per atom are sufficient to dissociate a N_2 molecule in diamond. Nevertheless, no NV-N center was found as long as T_{HB} = 300 K [14, 22].

After a N_2 beam was implanted into a diamond at $E_0 \leqq$ 500 eV/atom, the kinetic energy of a nitrogen atom, intrapair distance R_{N-N} and the LRO parameter were monitored for the first few ps using MD while the heat-bath was kept at T_{HB} = 300 K [14, 19]. The kinetic energy of either the implanted N atom or N_2 molecule damped in the order of thermal vibration. The molecule dissociated gradually over hundreds of fs and corresponding "R_{N-N}" expanded from 0.1 nm to a larger value with a slight dependence on the impact energy [14]. On the other hand, the global LRO parameter that integrated over all atoms showed a definite phase transition from a crystalline to an amorphous (CA) state, i.e., $LRO = 1$ to $LRO = 0$, after a few ps since starting the implanatation [19]. A similar CA transition was observed in the case of a low energy (~ 100 eV/atom) boron cluster (B_n; $n \leqq 6$) implanted in c-Si [23]. The very slow molecular dissociation was the result of multiple collisions due to small angle scattering [19]. We supposed this can be the reason why sub-keV/atom implantation is high enough to dissociate a N_2 molecule to an appropriate R_{N-N} distance. The very slow onset of the CA transition will be discussed in more detail in the next section.

3.2. Very Slow Response of the Damaged Crystal

Figure 1 shows a typical profile of the $LRO(z)$ how the radiation damage proceeds globally within the calculation volume in the first 5 ps, after a single N_2 molecule implantation into a diamond at T_{HB} = 300 K with energy of sub-keV (E_0 = 200 eV/atom).

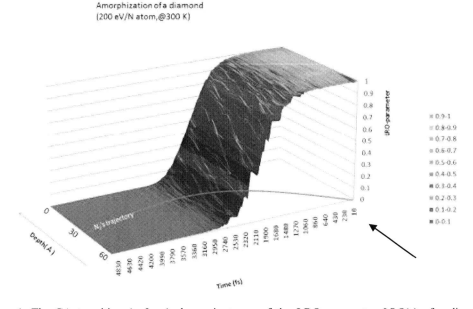

Figure 1. The CA transition (~ 2 ps) shown in terms of the LRO parameter $LRO(z)$ of a diamond bombarded by N_2 beam with 200 eV/atom at T_{HB} = 300 K. The arrow (~ 500 fs) indicates a stage where almost kinetic energy of N_2 was lost. The depth dependence was obtained by slicing the target into layers with a thickness of 0.5 nm.

The solid curved line indicates the mean depth of the two nitrogen atoms after impingement. The arrow (~ 500 fs) indicates a stage when the projectile almost lost its original kinetic energy as evidenced by the proximity to its asymptotic projected range, but where no appreciable change appeared in the LRO parameter. However, much later, at around ~ 2 ps, a definite change in the LRO parameter emerged that indicated the occurrence of the CA transition. The time-scale of this CA transition is about tens of the period of lattice vibration, which is reasonably long enough to stimulate global phonons in a damaged crystal. No significant depth dependence was observed in the $LRO(n_z)$.

No annealing of the host material nor the replacement of dopant was observed for T_{HB} = 300 K, even expanding when the computation time was extended to the order of 100 ps. When annealing was carried out, after elevating T_{HB} up to 1000 K, during the "*long black-out period*" an indication of crystal recovery was weakly evident before 800 ps. Also in the case of c-Si, during a similar annealing process, a black-out was observed although with a somewhat oscillatory behavior of recovery [23]. Note that the Debye temperature of a diamond (> 1860 K) is about three times larger than that of c-Si. These different reactions in the materials' response to irradiation may imply that preparation of recovery happens slowly during the long blackout period characterized by the long tail in Figure1.

3.3. Formation of NV-N Centers by Hot-Implantation

No NV-N center was formed for T_{HB} = 300 K. In the case of hot-implantation, we found the formation probability of the NV-N center was 3.7 % with R_{N-N} of 1.73 nm at 900 K, and 10.5 % with R_{N-N} of 1.46 nm at 1000 K [19, 22]. The identification of the substitutional N atoms was done by the PM Table [17, 18]. This result shows that the heating process helped to remove the crystal damage and to promote the replacement of a lattice atom with a dopant.

CONCLUSION

New material functions can be designed using ion implantation. Diamond is leading the emerging research materials in the ITRS [1]. For realizing a room temperature quantum computer (QC) nitrogen doping by hot implantation can produce a defect center made of a substitutional nitrogen and adjacent vacancy, i.e. the NV center. For the architecture of a QC, the arithmetic element and logic element are the key units, for which the NV and NV-N centers are currently the most probable candidates. The indispensable conditions are an appropriate separation between two N atoms (R_{N-N}) at around 2nm and both N atoms are on lattice sites. Making use of a molecular dynamic simulation (MD) and a crystallographic analysis called Pixel Mapping (PM) [17], we confirmed that sub-keV hot- implantation was high enough in energy to decompose a N₂ beam and position the two N atoms properly apart, i.e., without the need to adopt a few-keV/atom energies [11] to obtain an appropriate value of R_{N-N}.

We calculated long-range-order (LRO) parameter, because we supposed that the diamond structure will be retrieved after annealing. A global crystalline to amorphous (CA) phase transition of the target material occurred considerably later than the projectile had lost its incident energy. The CA transition [16, 19, 23] was clearly observed at around a few ps and with a long black-out period continued up to 800 ps before crystal recovery started. This indicates the annealing process needs a long elapsed time to replace the dopant properly at a lattice point in a well recovered crystal.

ACKNOWLEDGMENTS

The author thanks Professor H. J. Whitlow of University of Jyväskylä in Finland for clear and outstanding discussions for the interaction of ions with material, and Professor D. Neil of Okayama University of Science for encouraging discussions.

REFERENCES

[1] "International Technology Roadmap for Semiconductors (2009)", http://www.itrs.net/ links/ 2009 ITRS/Home2009.htm.

[2] D. G. Pettifor. *"Many-Atom Interactions in Solids"*; Eds. by R. M. Nieminen et al.; Springer Proceedings in Physics, 48, Springer-Verlag: Berlin, Heidelberg. 1990; pp.64-84.

[3] J. B. Gibson, A. N. Goland, M. Milgram, and G. H. Vineyard. *Phys. Rev.* 1960, 120, 1229-1253.

[4] W. D. Brenner, O. A. Shenderova, J. A. Harrison, S. J. Stuart, B. Ni, and S. B. Sinnott. *J. Phys. Condens. Matter.* 2002, 14, 783-802.

[5] *"Properties and growth of diamond"*, Ed. by G. Davis, (Inspec, IEE, 1994)

[6] L. A. Bursill and R. W. Glaisher. *American Mineralogist.* 1985, 70, 608-618.

[7] J. A. Larsson and P. Delaney. *Phys. Rev.* B, 2008, 77, 165201.

[8] E. van Oort, P. Stroomer, and M. Glasbeek. *Phys. Rev. B* 1990, 42, 8605-8608.

[9] W. K. Wootters. *Phys. Rev. Lett.* 1998, 80, 2245-2248.

[10] A. D. Greentree, N. A. Fairchild, F. M. Hossain, and S. Prawer. *Material Today* 2008, 11, 22-31.

[11] T. Gaebel, M. Domhan, I. Popa, C. Wittmann, P. Neumann, F. Jelezko, J. R. Rabeau, N. Starias, A. D. Greentree, S. Prawer, J. Meijer, J. Twamley, P. R. Hemmer, and J. Wrachtrup. *Nature Phys.* 2006, 2, 408-413.

[12] R. Hanson, F. M. Mendoza, R. J. Epstein, and D. D. Awschalon. *Phys. Rev. Lett.* 2006, 97, 087601.

[13] S. T. Nakagawa. *J. Surf. Sci. Soc. Jpn.* (in Japanese) (Special issue of the Cluster Beam and application) 2010, 31, 580-586.

[14] S. T. Nakagawa, H. Kanda, and G. Betz. *Trans. Elec. Elec. Eng.* 2010, C130, 2182-2187.

[15] S. T. Nakagawa. *Trans. Elec. Elec. Eng. (in Japanese)*, (Special issue of the "Nanobio science and fundamental technology") 2009, C129, 238-244. The English version appears at http://www3.interscience.wiley.com.

[16] S. T. Nakagawa. *Phys. Rev. B*, 2002, 66, 094103.

[17] S. T. Nakagawa. In *"Ion beams in Nanoscience and Technology"*; R. Hellborg, Editor and H. J. Whitlow, Editor, and Y. Zhang (Eds.); Springer-Verlag: Berlin, 2009; Chapter 9, 129-145.
[18] S. T. Nakagawa. *J. Phys. Soc. Japan* 2007, 76, No.3, 034603/1-12 (2007).
[19] S. T. Nakagawa, H. Nagao, and G. Betz. *Diamond & Related Materials,* 2011, 20, 927-930.
[20] S. T. Nakagawa, H. Hashimoto, H. Kanda, A. Okamoto, M. Ohishi, H. Saito, and G. Betz; *Nucl. Instr. Meth.* B, 2009, 267, 1226-1228.
[21] S. T. Nakagawa and G. Betz; *Nucl. Instr. Meth.* B, 2001, 180, 91-98.
[22] S. T. Nakagawa, H. Kanda, T. Sakai, M. Ohishi, H. Saito, S. Nakagawa, Y. Banden, and G. Betz. (2011) *Trans. Mater. Res. Soc. Jpn.*, 2011, 36, 79-82.
[23] S. T. Nakagawa (2011). Proc. of the 11th Int. *Workshop on junction Technology* (IWJT2011), 2011, 40-43.
[24] S. T. Nakagawa. *J. Surf. Sci. Soc. Jpn.* (Special issue of the "Cluster Beam and application"), 2009, 52, 224-230 (2009).

Reviewed by Dr. H. Kanda of NIMS *(National Institute for Materials Science) in Japan.*

Chapter 9

DIAMOND AND RELATED MATERIALS FOR BIOLOGICAL APPLICATIONS

Andrew Hopper and Frederik Claeyssens[*]
University of Sheffield, Materials Science and Engineering Department, Kroto Research Institute, United Kingdom

ABSTRACT

Diamond and related materials, such as diamond like carbon (DLC) are currently widely studied as materials for biology, both in thin film form (for substrates of biosensors or to enhance cell growth) and in nanoparticle form (as fluorescent markers and drug delivery vehicles). This review highlights and summarises important advances in the emerging field of diamond based biomaterials. Its beneficial electrical and chemical properties such as a high refractive index, low surface roughness, biocompatibility and corrosion-resistance make diamond and its related materials suitable for a wide range of biological applications. Additionally, reliable chemical functionalisation routes of diamond surfaces have recently emerged, enabling the controllable covalent attachment of biomolecules onto this surface.

Nanodiamond (ND) and ultrananocrystalline diamond (UNCD) films can be deposited under relatively mild conditions to biological surfaces as a substrate for further chemical functionalisation. Such chemical additions could pave the way for a variety of future applications such as in biosensing and microelectromechanical (MEMs) devices. Diamond films have also been suggested as a suitable method of improving the biocompatibility of objects which may come into contact with physiological fluids *in vivo* such as surgical tools and artificial prostheses.

At the smaller length scale diamond nanoparticles have also been synthesised and have been suggested as suitable carriers for drug and gene delivery, achieving promising results for insulin delivery into cells. Nanodiamond particles may also be utilising for bioimaging purposes, where their small size (5 nm diameter) enables them to cross the cell membrane permitting their observation within individual cells by fluorescence microscopy. Similarly, carbon nanodots (C-dots, carbon nanoparticles with diameters typically below 10 nm) are also well suited to roles in bioimaging. Taken up by cells

---

[*] Corresponding author: F.Claeyssens@sheffield.ac.uk, +44 (0)114 2225513

through endocytosis, they are thought to be potential successors to quantum dots which have limited applications for *in vivo* and *in vitro* imaging due to their inherent cytotoxicity. Their widespread advantages, such as low cost synthesis, chemical stability, ability for conjugation/functionalisation and resilience to photobleaching illustrate their potential for applications in optical imaging.

The remarkable properties exhibited by diamond and its associated forms continue to appeal to scientists as research in the area constantly proliferates. It is envisaged that diamond-based materials will play an increasingly important role in the future of biomaterial development.

1. INTRODUCTION

Diamond and its related amorphous carbon derivatives are currently being researched and considered for use in a variety of biological applications [1-16]. The many unique properties that diamond possesses are unparalleled in the natural world. The carbon-carbon distance (1.54Å) of bulk diamond crystal results in it possessing one of the highest atomic densities of any solid [4], being only surpassed by nanoaggregates of diamond nanorods [17]. Furthermore, the exceptionally high bond energy between the carbon atoms and the associated directionality of the respective tetrahedral bonds in the material's structure conveys diamond's impressive strength [18]. The unusually low coefficient of friction further adds to the list of tribological advantages which diamond holds and why research into its possible future applications continues to progress.

In addition, recent advances in the chemical surface functionalisation of diamond have enabled the covalent binding of biomolecules onto the diamond surface and have enabled exciting novel biomedical applications. Diamond can be rendered semiconducting via doping with boron and phosphorous which makes it to be well suited as a biosensing material; indeed important progress has recently been made in this area [4]. Diamond substrates have been functionalised with oligonucleotides to form biologically active DNA biosensors which exhibit greater stability and selectivity than those constructed from more conventional materials such as gold coated glass substrates [19]. The lack of cytotoxicity associated with diamond and amorphous carbon, such as DLC, also opens the possibility of them being utilised in more widespread *in vivo* applications. Already, nanocrystalline diamond has been suggested as a material for retinal prostheses [6, 20].

Temperature sensitive substrates (e.g. silicon or glass based devices) can be surface coated with an amorphous form of carbon (diamond-like carbon, DLC) which can produce a surface coating upon substrates under much milder regimes compared to diamond. Diamond-like carbon (DLC) can be fabricated under ambient conditions as a thin film (typically having a thickness 10-100 nm) via plasma coating, pulsed laser deposition or cathodic arc deposition, thus providing a low-temperature route to surface modification and chemical stability. DLC was initially used for its anti-wear and haemocompatible properties as a surface coating for orthopaedic joints and surgical instruments [21]. For example, DLC has been applied as a coating to the articulating surface of joint prostheses [22]. These surfaces can degrade over time, with the released wear particles causing inflammation. The application of a DLC coating to such areas creates a hard, wear-resistant layer, which would significantly increase the longevity of these implants and improve the quality of life for patients. Additionally, this coating can be surface functionalised via the same routes as diamond which can create, with relative ease, effective biosensors or biomolecule arrays.

A nanoparticulate form of diamond (nanodiamond, ND) has also been widely investigated for its possible biological applications. ND can exist as individual particles of diameter 5-10 nm or as thin surface films/monolayers. It has been illustrated that ND particles or films can be functionalised via covalent or non-covalent surface modification, both of which have yielded encouraging results. Enzymes such as lysozyme and luciferase have been attached to ND particles by non-covalent functionalisation, and have been shown to retain their catalytic activity following immobilisation. DNA strands can be grafted onto hydroxylated or oxidised ND films via covalent modification. ND particles have also been studied for their suitability as drug and gene carriers into living cells latterly, via coupling, with biological molecules such as antibodies. Of further interest, ND particles which have been doped with certain elements have been shown to luminesce due to the presence of point defects. Following nitrogen doping, for example, ND particles have fluorescence wavelengths in the far-red, between 638-780 nm. Such particles may then be introduced into cells, enabling them to be monitored via fluorescence microscopy. No discernible photobleaching or cytotoxicity have been known to occur following such a procedure, illustrating the advantages such particles possess over conventional fluorescence labels for imaging living cells.

Furthermore the widespread use of nanoparticles for biolabels and imaging purposes is becoming ever more widespread, with new and exciting areas emerging, such as the use of carbon nanodots for *in vivo* and *in vitro* bioimaging. These are thought to have great potential as safer alternatives to quantum dots which are produced from heavy metals such as cadmium. These metals are toxic and human carcinogens, exemplifying the need for suitable replacements as biological imaging markers [16]. Similarly fluorescent nanodiamonds (FNDs) can be individually detected in, and taken up by, cells and their movements monitored to track cellular metabolism pathways [23]. Recent work has uncovered the efficiency of diamond nanoparticles as drug carriers for insulin. The enzyme remained dormant whilst non-covalently bound to the nanodiamond particles, however, following release in an alkaline environment the enzyme regained it biological functionality. These finding could have widespread consequences to the future of drug delivery, allowing more comfortable and easier drug administration methods.

This review aims to provide an overview to the current biological research being conducted into diamond and amorphous carbon films and nanoparticles. In order to provide an understanding as to what applications may be available to these unique classes of materials we start this review with an overview of the various functionalisation routes which are applicable to diamond and amorphous carbon coatings.

2. FUNCTIONALISATION ROUTES

As indicated in the introduction, both DLC and diamond exhibit numerous advantageous properties. One important characteristic for their use as a biomaterial is the ability to chemically graft moieties and functionalities onto their surface via well-defined and easily accessible surface functionalisation routes. Through these procedures it is possible to optimally exploit the bulk properties of diamond and its related materials in biomedical applications, enabling the manufacture of diamond based biosensors, gene and drug delivery vehicles and biolabelling particles. The functionalisation of these materials also chemically homogenises the surface enabling further modification to occur more easily and effectively. Such functionalisation of amorphous carbon and diamond-based materials can take place either through covalent or non-covalent means producing surfaces which have a variety of uses for biological purposes as mentioned later in this review.

2.1. Non-Covalent Functionalisation

Although amorphous carbon surfaces prove to be particularly inert under ambient conditions the adsorption of biological molecules, such as proteins [24-28], can nevertheless occur. This is true for oxidised amorphous carbon surfaces, whose terminal groups (hydroxide, aldehyde/ketone and carboxylate moieties) are able to participate in polar interactions, such as hydrogen bonds, to bind with enzymes such as luciferase [28] and lysozyme [29]. These functionalised surfaces were shown to be biologically active, with the activity of the adsorbed enzymes being conserved. It is possible to attach further molecules on to the functionalised surface as demonstrated by Huang et al. [25]. Their research illustrated that it was possible to covalently immobilise fluorescent dyes to poly-L-lysine coated nanodiamond particles, allowing them to be utilised for labelling procedures (see Figure 1).

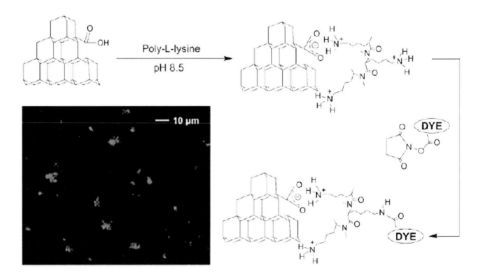

Figure 1. Non-covalent surface modification of nanodiamond particles using Poly-L-Lysine (PLL). After electrostatic binding of the PLL a fluorescent dye was covalently bound to the amino group of the Poly-L-Lysine. Inset is a fluorescence image of the nanodiamonds labelled with Alexa Fluor 488. Reproduced with permission from [25].

Nanodiamond has also shown the ability to non-covalently bind with antibodies [30] and peptides [31]. Wang *et al.* [32] confirmed the suitability of nanodiamond as a protein carrier via non-covalent adsorption. This was illustrated using the example of bovine serum albumin (BSA) which is known as a 'soft' protein due to its low internal stability. This translates to it being one of the most susceptible proteins to undergo conformational alteration following environmental change. Although slight conformational changes to the protein structure were observed following immobilisation, these are unlikely to result in protein denaturation. The nanodiamond was shown to preserve most of the BSA structural features and exhibited a high affinity for adsorption.

2.2. Covalent Functionalisation

A plethora of covalent functionalisation strategies exist for the surface modification of carbon substrates, both diamond and amorphous, with research continuously expanding [33, 34]. One set of methods is based on photochemical surface functionalisation. This method utilises ultraviolet radiation (λ ranging from 254 nm to 360 nm) to facilitate the binding of organic molecules possessing a terminal vinyl functional group to a hydrogenated carbon surface. The stable C-C bond that is formed via the addition reaction of these alkenes to the surface create a base on which functional moieties can be attached. The advantage of using photochemical techniques is that arrays of functional groups can be precisely produced through simple shadow mask techniques [35].

The photochemical reaction involving organic alkenes is initiated by ejection of a photoelectron from the sample. This creates radicals in the solution which cleave hydrogen atoms from the diamond surface, creating highly reactive sites that react with the alkenes [36]. One such alkene which has particular interest to biological research is trifluoroacetamide- protected 10-aminodec-1-ene (TFAAD). Following adsorption, and the deprotection of the TFAAD molecule, a free amine functional group is yielded onto which further biological moieties such as DNA or enzymes can be bound to form highly sensitive and stable biosensors [37]. Additional chemical methods of functionalising hydrogenated diamond films include the electrochemical or radical grafting of diazonium salts and azo-perfluoroalkyl compounds respectively. Further information on these procedures has been detailed in a number of studies [38-42].

In contrast with diamond and diamond-like-carbon deposited thin films, diamond nanoparticles produced by detonation methods are initially difficult to surface modify without chemical refinement. The chemical inhomogeneity of detonation diamond nanoparticles is one of the first stumbling blocks towards effective surface functionalisation. The particles are coated in soot, graphitic carbon and metal impurities causing the particles to stick together into micrometre-sized aggregates (see Figure 2). These impurities can be removed by oxidising the nanodiamond in an acid solution, concurrently causing deaggregation. It is then desirable to chemically homogenise the nanoparticles through a variety of different functional routes. One such method utilises reactive gases, including ammonia, hydrogen, chlorine and carbon tetrachloride [43] (see Figure 3). This treatment, whilst improving the purity of the nanoparticles, also established homogeneous functional groups (NH_2, C-H and Cl) on their surfaces. Further research established that chlorinated nanodiamonds could undergo amination following exposure to ammonia [44]. Similar results were obtained using fluorine in the presence of hydrogen [45]. The fluorinated surface coating could then be further functionalised through the binding of amino acids or amines.

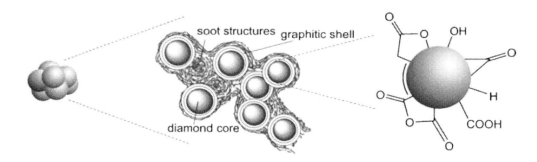

Figure 2. Schematic representation of the structure of aggregated detonation diamond illustrating the diamond particles embedded in a graphitic matrix. The individual diamond particles exhibit a number of different surface functional groups as illustrated on the right. Reproduced with permission from [34].

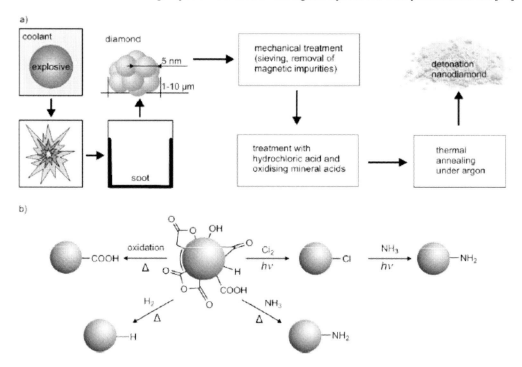

Figure 3. (a) Schematic illustrating the production and purification of detonation nanodiamond and its covalent functionalisation by oxidation or solid-phase reaction with chlorine, ammonia or hydrogen. Reproduced with permission from [34].

In addition to the use of reactive gases, more easily accessible wet chemistry routes exist for the chemical surface homogenisation of diamond nanoparticles. The surface of the oxidised nanodiamond powder consists of a mixture of hydroxyl, anhydride, carboxyl (ester and acid) and carbonyl (anhydride and ketone) moieties (illustrated in Figure 2). These surface moieties (except for hydroxyl groups) can be reduced with borane to produce a uniformly hydroxylated surface. These hydroxyl groups, as reported by Krueger *et al.* allow for further modification by the grafting of trialkoxysilane molecules [33]. Once adsorbed, the amino groups of these silane molecules, such as trimethoxy- and triethoxy-3-amino-propyl-silane, can be used for further surface modification. The successful adsorption of biotin through this process was accomplished, with its biological activity maintained to allow for its binding to streptavidin [46]. Krueger has also conducted research into the beneficial procedure of functionalising nanodiamond particles using 'click chemistry' [47]. This involves the copper(I) catalysed cycloaddition of azides and alkynes, to which additional organic moieties can be bound or 'clicked'. This can enable the multifunctionality of diamond nanoparticles, leading them to exhibit both carboxylic and alkyne functional group. Such a method permits the concomitant binding of two biomolecules at different binding sites through orthogonal coupling linkers. This procedure is biologically significant since the technique is highly efficient and selective, possibly allowing for nanodiamond particles to be utilised as multifunctional biolabels and drug delivery vehicles [47].

A powerfully alternative method for producing hydroxylated nanodiamond particles directly from detonation diamond has been devised by Martin *et al*. This procedure uses fenton chemistry to significantly increase the density of hydroxyl groups upon detonation nanodiamond, schematically illustrated in Figure 4 [12]. Fenton reactions are known to produce highly reactive hydroxyl radicals that can cause partial oxidation of any exposed carbon atoms on the nanodiamond. Concurrently, the reaction also causes the degradation of any absorbed soot on the surface, forming carbon dioxide, thus exposing greater numbers of carbon atoms for oxidation, forming a surface layer of –OH groups onto which a high density network of functional groups can be covalently bound. X-ray diffraction (XRD) of the fenton treated nanoparticles clearly revealed the diamond XRD peaks, indicating that the method preserves the diamond crystal structure whilst increasing functionality.

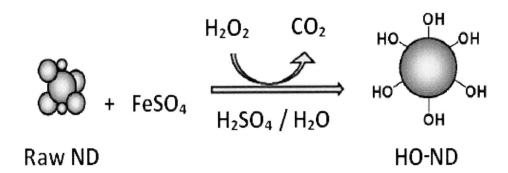

Figure 4. Schematic illustrating the fenton reaction to functionalise nanodiamond particles. Reproduced in altered form with Permission from [12].

Additionally, it is possible to functionalise nanodiamond with aryl organic molecules by Suzuki coupling reactions [48]. This reaction route is useful for producing fluorescent particles, through the Suzuki coupling of pyrene, which generates a fluorescent nanodiamond-pyrene complex. Importantly, nanodiamond functionalised in this manner demonstrates greater resistance to aggregation, and improved dispersion in ethanol and hexane. The functionalisation strategies in this section illustrate the range of possible applications which diamond and related carbon materials are suited for. Research continues to progress in identifying further functionalisation methods and applications which could prove useful for future research. Past and present biological uses of diamond related materials will now be discussed in the following sections.

3. NANODIAMOND FILMS

Another form of diamond which has been shown to have uses in biological research, as well as in electrical applications is nanocrystalline (NCD) diamond. Discovered in the early 1960s in the Soviet Union [49], the material can be obtained as a film by chemical vapour techniques (CVD) (see Figure 5) [50] and can be distinguished from other forms of diamond by its grain size, which is usually in the range of 100 – 300 nm. The growth mechanism for NCD films requires hydrogen-rich growth chemistry, whereby crystal growth results in a columnar morphology which is oriented in the growth direction. As the film increases in thickness, the grain sizes subsequently increase also. It is possible to maintain grain sizes in the NCD range for films several microns in thickness by controlling the nucleation density and growth parameters [4].

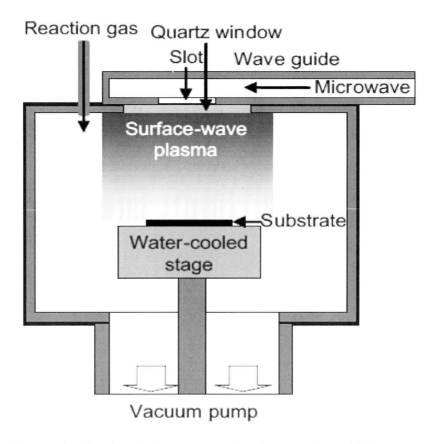

Figure 5. Diagram depicting the typical components of a microwave plasma CVD system used for the low temperature deposition of nanocrystalline diamond films. Reproduced with permission from [50].

The desirable physical and chemical properties associated with NCD match those of natural diamond such as a negative electron affinity, wide band gap, optical transparency, exceptional hardness and remarkable biocompatibility. Such advantages have garnered accelerating research interest, opening the door to a possibly wide range of suitable applications. However, fabrication techniques such as CVD, which can often require temperatures in excess of 400°C for the production of NCD, may limit the range of substrates to which it could be applied [51].

Efforts have therefore been made to attempt the low temperature synthesis of NCD films. Hot-filament CVD (HFCVD) can deposit diamond particles on a silicon substrate at a temperature of 135°C [52], whereas microwave plasma CVD (MWPCVD) can fabricate NCD films at temperatures as low as 86°C on plastic substrates [50]. At such low temperatures, the crystal size of the diamond was measured on average to be 5 nm which is in keeping with normal NCD films.

However, in the past decade a new form of nanodiamond film, known as ultrananocrystalline diamond (UNCD) has been developed. It was patented and first produced at the Argonne National Laboratory [53] and has since been suggested as having possible uses in biological applications [6, 54]. UNCD possesses a much finer grain size, in the region of between 3-5 nm [55], in comparison to nanocrystalline diamond, whilst still sub-micron, having grain sizes which are significantly larger at between 100-300 nm [56]. As such, UNCD films are endowed with a larger surface area for the grafting of functional molecules, and a smoother surface topography which could potentially permit a greater cellular surface contact area [56]. Such films can be produced using a plasma which has a comparatively low hydrogen content, typically consisting of 97% argon, 2% hydrogen and 1% methane [57]. This is in contrast to the hydrogen-rich plasmas used for the synthesis of conventional nanodiamond films (99% hydrogen and 1% methane). It is possible to produce UNCD films of any thickness with little variation in the surface roughness due to a high re-nucleation rate [55]. In contrast, as the thickness of NCD is increased the surface roughness increases also, and the material will eventually resemble microcrystalline diamond [55]. It has been discovered that the bias-enhanced nucleation process for UNCD synthesis offers greater efficiency and enhanced substrate adhesion compared to other processes producing UNCD such as mechanical polishing. Chen *et al.* has developed such a technique that permits the deposition of UNCD layers of uniform grain size (3-5 nm), low compressive stress, ultra-smooth surfaces and high growth rates which could progress UNCD as a potential material for microelectromechanical (MEMs) devices [58].

3.1. Nanodiamond Films as Biocoatings

Research into the uses of NCD and UNCD as biomaterials have yielded a number of interesting results and applications, as illustrated by the following studies:

3.1.1. Surfaces Which Influence the Proliferation or Differentiation of Cells

Being able to control the surface properties of substrates for cell culture is of great scientific importance. Cell adhesion molecules, such as integrins, serve as interconnecting bridges between cells and the extracellular matrix (ECM) or cellular growth substrate [59, 60]. These integrins trigger signalling cascades, as illustrated in Figure 6, which determine cellular phenotype and differentiation capabilities. It has been demonstrated that the nanoroughness and surface charge distribution of a substrate can determine cell attachment efficiency, their proliferation and subsequent differentiation [61].

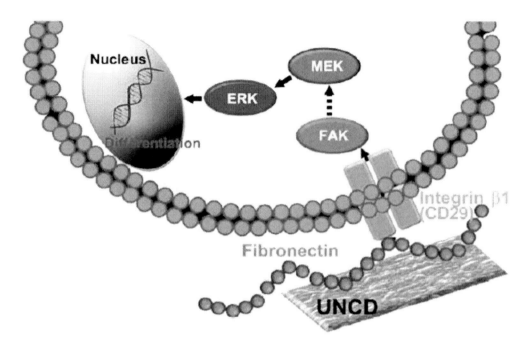

Figure 6. Schematic diagram summarising the influence of hydrogenated ultra-nanocrystalline diamond films (UNCD) in mediating neuronal differentiation from NSCs. Adsorbed fibronectin on H-UNCD activates integrin, Fak and Erk1/2 pathways culminating in ultimate and specific NSC neuronal differentiation. Reproduced with permission from [54].

Consequently, it would be advantageous to precisely control the differentiation and proliferation of stem cells *in vivo*. This is particularly true in the case of neural stem cells (NSCs) since injuries to the peripheral or central nervous system originate from the loss or damage of particular sub-populations of neural lineages [62, 63]. Neural stem cells cultured on UNCD have been found to differentiate into different cell types depending on the surface termination of the film [64]. NSCs cultured on oxygen terminated UNCD films, in serum-free media, preferred to differentiate towards oligodendrocytes (cells which insulate axons in the central nervous system (CNS)), whilst those cells cultured on hydrogen-terminated UNCD films were more likely to differentiate towards neuronal cells. These different responses seem to be directly related to the surface properties which both of these films exhibit. Hydrogenated (U)NCD films show negative electron affinity, possess p-type surface conductivity and are hydrophobic. Oxygenated (U)NCD films, on the other hand, are electrically insulating, hydrophilic and display a positive electron affinity [65]. These changes in cellular behaviour were quantified by reverse-transcriptase polymerase chain reaction (RT-PCR) through the expression of specific neuronal cell markers and immunofluorescence staining. The degree of differentiation was enhanced, compared to a polystyrene control, following the addition of neuronal growth factors. It was thought this is due to the composition of the (U)NCD film exerting an effect upon the local concentration of inherent growth factors, thus influencing the interaction of the adhering stem cells [64].

Hydrogenated UNCD films have also been shown to favour the adhesion and proliferation of mesenchymal stem cells, whereas oxygen and fluorine terminated UNCD coatings resisted cellular adhesion [66]. Oxygenated UNCD films have proved to be biocompatible for the growth of murine embryonic fibroblasts [67], whose proliferation was observed to be of a similar rate on both tissue culture plastic (TCP) and UNCD surfaces. Similarly no cytotoxicity was observed when fibroblast cells were cultured on NCD-coated ceramics, proving the substrate to be a suitable surface for cell attachment and growth. After a few days the cells were observed to form a confluent surface layer, with proliferation rates observed to be slightly higher than those observed on TCP [68].

Cell lines have been shown to favour oxygenated-(U)NCD (hydrophilic) surfaces, since this provides a plethora of interaction sites for media components and adhering cells. Such sites are the result of local variations in electric field gradients across the (U)NCD film, caused by oxygen termination, polarity and surface alterations. Cellular adhesion was significantly greater than that recorded on untreated glass. In contrast, it was illustrated that hydrogenated (U)NCD films did not provide sufficient interaction sites for satisfactory adhesion for cell-line cells [65].

However, UNCD films can also be functionalised in such a way as to reduce cellular activity [69]. It is possible to coat carboxylated-UNCD with collagen to which dexamethasone (Dex), an anti-inflammatory therapeutic, can be bound. Dex is a glucocorticoid which can attenuate inflammatory gene expression and could have useful applications as a surface coating to artificial implants and materials to reduce potential inflammation and biofouling. The presence of Dex has been shown to reduce the incidence of pro-inflammatory cytokines such as IL-6 and suppress inflammatory gene expression.

3.1.2. Biosensors

One major hurdle for the integration of microelectronics and biotechnology is the requirement to develop interfaces that are compatible with current processing methods, whilst also permitting satisfactory selectivity and *in vivo* stability. Conventional microelectronic materials including silicon and gold can be biologically functionalised, but surface degradation can be problematic, affecting the efficiency of the associated biosensor. Attention has therefore shifted to the use of diamond, in particular NCD, for the construction of integrated biosensors. Its aforementioned chemical and physical properties, in addition to its ability to be deposited as a thin, robust film, prove desirable for biosensor construction. Indeed Yang *et al.* [19] were able to attach DNA molecules to hydrogenated NCD films which had been photochemically functionalised with TFAAD. The resulting biosensor was shown to have excellent selectivity and stability which exceeded that of conventional microelectronic compatible substrates such as silicon, gold and glass [19]. The potential for using U(NCD) as a biosensor platform has been illustrated by the study of Härtl *et al.* [70]. This work reported the production of a hydrogen peroxide sensor based on catalase being covalently bound onto a NCD substrate. These biosensors showed excellent activity, indicating low denaturation of the enzyme upon binding to the NCD surface.

It has also been discovered that diamond surfaces functionalised with ethylene glycol groups (EG) discourage the non-specific binding of proteins. Stavis *et al.* [71] demonstrated that such a system reduced the detection of adsorbed fibrinogen beyond the detection range capability of XPS, experiencing a reduction in adsorption of at least 97% in comparison to hydrogenated-diamond samples. Such data accounts for the increased signal:noise ratio evident from biosensors created from diamond substrates in comparison to those manufactured upon gold or glass. Encouragingly, it was discovered that the roughness of the diamond surface (nanocrystalline diamond films exhibiting greater roughness than single crystal diamonds) had little effect upon covalent binding ability. Furthermore, the binding of *E.Coli* antibodies to diamond thin films was demonstrated to be particularly stable. Glass substrates experience difficulty in maintaining the covalent bond formed with the antibody due to hydrolysis of Si-O-Si bonds at the antibody interface. The biological activity of the attached antibodies on diamond surfaces was also better than that recorded on glass, illustrating that biosensors created from diamond could have greater longevity [71]. Such findings, relating to the increased stability and biological activity of adsorbed antibodies, was further corroborated by Radadia *et al.* [72] who reported that antibody-derived biosensors could undergo repeated antigen-binding cycles without any significant loss in activity. Such sensors may have possible applications such as the monitoring of food and water supplies for contaminating bacteria including salmonella, E. Coli and Campylobacter.

Immunosensors have also been developed, such as the one produced by Bijnens *et al.*, using nanocrystalline diamond [73]. This sensor was capable of physically adsorbing anti-C reactive protein antibodies to a hydrogenated NCD film surface. It was possible to monitor these proteins in real-time, permitting the detection of micromolar quantities within a timeframe of 30 minutes.

3.1.3. Retinal and Neural Prostheses

Retinal microchips may enjoy widespread future use in restoring a patient's vision following retinal degeneration due to diseases such as retinitis pigmentosa or age-related macular degeneration. The idea would be to implant an optical microchip within the eye of the patient to replace the photoreceptors and connect this device to the inner retina [6, 20, 74]. For such an implant to be successful, the prosthesis material needs to be constructed of, or coated with, a bioinert material to prevent inflammation of the retina or the release of toxic or irritable by-products. It was discovered that although silicon would be suitable as a structural and functional material for the microchip, its unfortunate ability to slowly dissolve in physiological fluids required the use of a coating material, such as (U)NCD, to protect the chip's integrity. This coating procedure was explored by Xiao *et al.* [6]. Deposition of NCD films by PECVD at temperatures between 400 - 800°C did not adversely affect the Si substrate. Implantation of these (U)NCD coated Si chips into the eyes of rabbits over a 6 month period elicited no immune response, confirming the bioinertness and biostability of (U)NCD coatings and their future importance to this biological application [6].

Ganesan *et al.* [20] have detailed research into the production of a diamond-penetrating micro electrode array (MEA) which could in future restore at least partial vision to patients without eyesight. It was possible to implant a model of the array into rat retinas without causing significant damage to the diamond electrodes (95% of which survived the procedure intact). In a model retinal prosthesis the diamond electrodes are embedded in the patient's retina, residing in the ganglion cell layer in close proximity to the optic nerve. Images which are captured by a remote video camera can then be processed and relayed wirelessly to a microchip within the retinal device. The electrodes which would penetrate the retina could then electrically stimulate the ganglion cells, which in turn would then stimulate the optic nerve. Retinal cells have been shown to survive and proliferate on NCD equally as well as on glass [74] and it has been possible to fabricate diamond MEAs upon soft substrate materials such as polyimide and parylene to produce flexible implants. Following implantation in blind rats, no gliosis was observed in the retinal tissue, proving that the implant did not damage the optic nerve [74].

Similarly, Chan *et al.* [75] have published findings relating to the production of polycrystalline-diamond (poly-C) based microprobes, which may have future applications in neural prostheses. Boron-doped diamond was used as the electrode material providing a stable surface for the detection of chemical and electrical stimuli, whilst the supporting structure was composed of undoped polycrystalline diamond. It was possible to detect levels of norepinephrine, a neurotransmitter which acts as a stress hormone, at concentrations as low as 10 nM. *In vivo* studies, whereby the microprobe was inserted into the auditory cortex of a guinea pig brain to monitor neural activity were also carried out. Broadband sound signals were applied to simulate the brain, and neural activity within the cortex was monitored. Although the probe was capable of recording the stimulated neural signal, the signal to noise ratio was low. This may be due to the large surface area of poly-C, which is ~5 times greater than that of smooth metals, which can enhance background noise.

In conclusion, nanodiamond films have been researched in numerous applications in biotechnology. Its biocompatibility and electrical properties, whilst being able to be deposited as a coherent film, enable it to be of particular use as a biosensor substrate. Indeed the work which has been completed already with respect to retinal prostheses illustrates the future potential such a material can unlock.

4. NANODIAMOND PARTICLES

Industrial quantities of nanodiamond particles are mainly obtained by the high pressure shock-wave transformation of graphite and the other, more widely used procedure, is the oxygen-deprived detonation of explosives such as trinitrotoluene (TNT) or hexogen within a sealed steel container [34]. Such an explosion causes the synthesis of a fine powder, the particles having a diameter of approximately 5 nm. By mass, up to 80% of the resulting powder is diamond. Although this method can ensure the speedy synthesis of nanodiamond on an industrial scale, the extreme environment within which the detonation occurs can result in the deposition of an amorphous carbon layer, or the undesirable functionalisation of the particles' surface. The inherently high temperatures associated with the technique can also lead to the creation of a graphitic surface layer. However, such particles can be made chemically active through the application of the aforementioned surface treatments.

One of the main advantages of utilising nanodiamond particles is their ability for bio-conjugation. Even without chemical modification, the bare surface of nanodiamond particles is fairly reactive and will readily adsorb numerous types of molecules, including water. However following the surface chemical treatments previously mentioned a much greater number of biologically useful molecules can be readily bound to the nanodiamond surface, whilst retaining the extensive and favourable properties of bulk diamond. These include its extreme hardness, chemical stability and biocompatibility.

Recent work, however, has demonstrated that the presence of diamond nanoparticles can lead to slight increases in the expression of DNA repair proteins, such as p53 and MOGG-1, in murine embryonic stem cells [76]. This effect was found to be more pronounced in oxidised nanodiamond, indicating a possible surface chemistry-specific toxicity. It must be stated however that the expression of DNA repair proteins in this instance was significantly lower than those reported following exposure to multi-walled carbon nanotubes (MWNTs) [77]. Furthermore, the presence of additional surface moieties to create functionalised nanodiamond may counteract this perceived low level of toxicity, since this effect has not been witnessed in any further studies [78, 79]. On the contrary, a similar earlier study focusing on the effect of ND particles in embryonic fibroblasts found no discernible change in cellular gene expression [78]. Similarly, another paper further confirms the biocompatibility and lack of toxicity recorded after cellular exposure to raw and surface modified ND particles [79].

4.1. Applications of Nanodiamond Particles

At the present, nanodiamond particles have a number of applications in industry, mainly as being used in cutting tools and drill bits to create cutting edges that are smooth yet also extremely hard wearing, potentially increase the lifetime of a cutting appliance up to 100-fold [80]. Further applications include electroplating baths [81], cooling fluids [82] and lubricants [83]. Diamond nanoparticles have indicated to be highly interesting for biological applications, as illustrated by the following examples of published research:

4.1.1. Biolabelling

The ability with which biological systems function at the local level can be readily monitored through biolabelling and biomolecular probes and observing their interactions *in vivo*. Laser-induced fluorescence is often selected for this procedure due to its inherent sensitivity and flexibility in coping with simultaneous probe detection [84]. To avoid interference with endogenous cellular components, it is beneficial for biological probes to absorb radiation at a wavelength greater than 500 nm whilst fluorescing above 600 nm. Although organic dyes and fluorescent proteins can fulfil these obligations, inherent disadvantages, such as photobleaching, limit their potential uses for long term or prolonged *in vivo* research [84]. In contrast, nitrogen-vacancy point defects (N-V) in diamond emit fluorescence at approximately 700 nm, whilst absorbing strongly in the region of 560 nm. They are produced from the thermal annealing of irradiated diamond, and have been aptly termed fluorescent nanodiamonds (FNDs) [85]. FNDs exhibit fluorescence intensities similar to those of quantum dots and crucially, do not suffer from photobleaching. This important quality may mean that FNDs could be suited for long term observation in biological cells as single biomarkers, being taken up by cells through endocytosis [85]. This was proven by Fu *et al.* [84], who demonstrated it was possible to track a single FND in a HeLa cell as shown in Figure 7. More recent work has focused on conjugated FNDs for the uses of homogeneous labelling and superresolution imaging [23]. Albumin conjugated FNDs can readily couple with biotin, forming a biotin-albumin complex which can be further conjugated with avidin or streptavadin. Subsequently this superconjugate could bind with biotinylated antibodies for highly specific binding and imaging purposes.

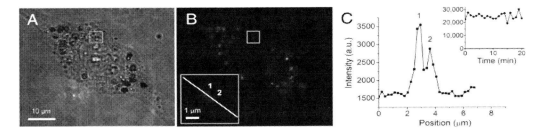

Figure 7. Observation of fluorescent nanodiamonds (FNDs) in a HeLa ell. Bright field (A) and epifluorescence (B) images of HeLa cell following uptake of 35 nm FNDs, with most nanoparticles residing in the cytoplasm (enlarged view of two FNDs shown in inset). The fluorescence intensity profile of the line drawn in Image B between FND 1 and 2 is also shown (C). Smaller graph illustrates fluorescence intensity from the sample after continuous excitation over a 20 minute period, showing no sign of photobleaching. Reproduced with permission from [84].

Additionally, it has been demonstrated that surface-carboxylated nanodiamond particles have been successfully used as in vitro probes in human lung cells (A549 lung epithelial and HFL-1 lung fibroblast cells), exhibiting natural green fluorescence by laser excitation which can be detected either via confocal microscopy or flow cytometry [16]. Industrially supplied particles of either 5 nm or 100 nm diameter were shown to be fluorescently active, with further research indicating that no discernible photobleaching occurs during analysis [85]. Additionally, the treatment of the lung cells with carboxylated nanodiamond particles did not induce significant cell death, apoptosis or altered protein expression [16]. Separate studies have confirmed that nanodiamond particles were non-toxic in human kidney cells [85] and neuroblastoma cells [79] further exemplifying the notion that diamond-based nanoparticles are biocompatible and suitable for future biomedical applications.

Of further note is the progress made by Zhang *et al.* [5] in producing a nanodiamond enriched poly-L-lactic acid (PLLA) scaffold for bone tissue engineering. The nanodiamond particles, functionalised with octadecylamine, when infused in the polymer to a concentration of 10 wt. % resulted in a 200 % increase in the polymer's Young's modulus and a concurrent 800 % increase in hardness. At these levels the polymer's properties approached that of human cortical bone. Furthermore the incubation of murine osteoblast cells within the polymer matrix over a one week period progressed without any noticeable toxicity or effect upon cellular proliferation. What is particularly interesting is the multifunctional potential of octadecylamine bound ND, since it can be further functionalised to perform as a drug delivery vehicle or as a fluorescent marker. Indeed, in this particular case the polymer matrix exhibited intense fluorescence allowing for the progress of *in vivo* bone re-growth into the implant to be monitored easily in the future [5].

4.1.2. Drug Delivery

Recently published data has raised the possibility of using nanodiamond particles as delivery vehicles for cancer drugs. The incidence of chemoresistant growths such as recurring mammory tumours and liver cancer has exemplified the need for an alternative, yet effective method of controlling this disease. In a study conducted by Chow *et al.* [2] administration of doxorubicin, an anthracycline antibiotic, bound to nanodiamond particles was tested in mice models of liver and mammary cancer. Results showed significantly increased apoptosis and tumour growth inhibition in the nanodiamond bound drug study in comparison to conventional doxorubicin treatment, by intravenous injection. Furthermore the *in vivo* ND-administration of the drug significantly decreased its toxicity compared to the conventional delivery method. It is therefore apparent that the nanodiamond bound drug conjugate could have increased efficacy whilst decreasing the severe side effects of the drug, the most serious of which are heart attacks. Earlier research conducted by Huang *et al.* have confirmed these results [86].

Furthermore, bovine insulin has been successfully bound, non-covalently, to diamond nanoparticles via electrostatic interactions as shown in Figure 8 [1]. This was confirmed by fourier transform infrared (FT-IR) spectroscopy and zeta potential measurements. The conjugated nanodiamonds exhibited pH-dependent desorption of insulin in sodium hydroxide solution. Adsorption and desorption could also be imaged by transmission electron microscopy (TEM) and quantified by the degree of protein functionality. The protein functionality was assessed by studying the insulin-mediated cell recovery of the RAW 264.7 cell line (demonstrated by 3-(4,5-Dimethylthiazol-2-Yl)-2,5-Diphenyltetrazolium Bromide (MTT) assay). Additionally, an RT-PCR assay indicated insulin-induced upregulation of Insulin-1 and granulocyte colony-stimulating factor gene expression in 3T3-L1 cells. These measurements indicated that insulin remained inactive whilst bound to the nanodiamond particles, but it was discovered that its biological activity had been preserved following desorption, illustrating the viability of the technique [1].

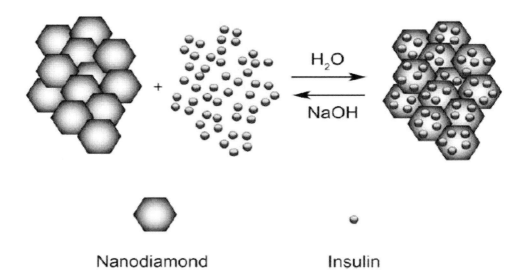

Figure 8. Schematic Illustration depicting insulin adsorption to ND particles in water and desorption in the presence of NaOH. Insulin binds non-covalently to ND in water due to electrostatic interactions. Alkaline conditions alter the insulin surface charge characteristics, releasing them from the ND surface. Reproduced with permission from [1].

4.1.3. Gene Delivery

The inherent lack of cytotoxicity in diamond nanoparticles, robust physical properties, drug carrying capabilities [1] and potential to migrate across cell membranes [12, 13] enable them to be ideally suited as carrier vehicles. Although a number of different methods already exist for the facilitation of nuclear material into nuclei, such as viruses, charged lipids and cationic materials, the extremely low cytotoxicity of diamond nanoparticles is an essential property to justify future *in vivo* research. The lack of patient confidence in virus-mediated gene therapy, along with related toxicity, immune and inflammatory issues [87] illustrates the excellent potential and advantages diamond nanoparticles could play in gene therapy.

It has recently been reported [12] that fenton-treated diamond nanoparticles could be successfully functionalised with thionine and injected into culture medium where HeLa cells were residing. Thionine emits red fluorescence upon excitation at 598 nm, a unique characteristic which can be utilised to locate the position of thionine-bound objects. After 24 hours incubation in the nanoparticle enriched media, fluorescence microscopy illustrated the presence of the functionalised nanodiamond within the cell nuclei. Quantitative flow cytometry analysis indicated that cell toxicity did not deviate from the control during incubation, confirming the biocompatibility of nanodiamond particles in vivo. Cell nuclei also demonstrated normal morphology, showing no signs of apoptosis.

Further research resulting in the migration of plasmid-bound diamond nanoparticles into the nuclei was demonstrated. The oxidised nanodiamond particles were functionalised with a triethylammonium salt onto which a plasmid incorporating the green fluorescent protein gene (GFP). It was demonstrated that the plasmid alone could not enter the cell or nucleus of its own accord. Upon migration into the cell nucleus the GFP gene was expressed and transcribed, proven by fluorescence microscopy (excitation at 488 nm) and Western blotting [12].

4.1.4. Intracellular Antioxidant System

Detonation diamond, treated under Fenton conditions, displays large surface concentrations of hydroxyl groups which are able to support gold and platinum nanoparticles (diameter: 2 nm). Gold and platinum are able to trap organic radicals and furthermore display high peroxide catalytic activity, enabling them to act as antioxidants. The ability of diamond nanoparticles to cross the cellular membrane enables them, when coupled with gold or platinum particles, to act as reactive oxygen species (ROS) quenchers, reducing cellular oxidative stress. This was confirmed by research conducted by Martin *et al.* [13], whereby levels of ROS were significantly reduced following administration of rotenone, a mitochondrial inhibitor, to Hep3B cells. Interestingly, the degree of ROS decomposition was greater than that of the conventional antioxidant glutathione, with the most beneficial results being obtained from gold nanoparticles [13].

4.1.5. Influence Upon Cell Attachment

The effect of ND substrates on the attachment and proliferation of primary murine neuronal cultures has also been studied and produced interesting data [7]. It was established that that the addition of ND particles on glass, mechanically polished polycrystalline diamond (PCD), nanocrystalline diamond (NCD) and silicon (Si) promoted the attachment of primary murine neurons. Neuronal attachment was more or less absent on these same substrates if they were not coated with ND. Furthermore, ND-coated materials exhibited similar neural attachment and outgrowth as displayed on more conventional ECM-derived substrates such as laminin, and were capable of supporting both neuron and glia cells. This suggests that ND based substrates may have future applications in the development of specialised neuronal growth surfaces.

To conclude, research conducted to date has shown the wide range of applications for which diamond nanoparticles appear to be suited. Its lack of toxicity enables it to be extremely suitable for biological research, having applications such as fluorescent labelling; vehicles for genes, drugs and antibodies; and the adsorptive separation, purification and analysis of proteins [34]. The ability to graft functional molecules including enzymes and DNA onto the nanodiamond surface also is particularly useful, and is partly the reason why in the future it is anticipated that nanodiamond will play an increasingly important role in biological research.

5. DIAMOND-LIKE CARBON (DLC) AS BIOCOATINGS

Diamond-like carbon (DLC) is an amorphous form of carbon which is composed of a network of sp^3 and sp^2 hybridised carbon atoms. The ratio of these atoms within the DLC is partly responsible for the properties of the deposited film. However, the concentration of hydrogen within DLC also influences its properties; the greater the concentration of hydrogen, the more ductile, or polymeric, the DLC becomes. It is these properties which can have an influence upon which type of DLC is required for a specific biological application and determines the different forms of DLC which are available as depicted in the phase diagram of Figure 9.

5.1. Different Forms of DLC

Many different categories of DLC have been reported. Any such film which has been produced can be grouped into one of three categories, dependent upon their hydrogen content and the percentage of sp^3 carbon.

5.1.1. Polymeric (Highly Hydrogenated) Amorphous Carbon (a-C:H)

DLCs (type a-C:H) which have a composition greater than 40 at. % hydrogen are termed polymeric amorphous carbon. They have comparatively low hardness compared to the other forms of DLC listed below and are ductile. Both of these properties can be explained by the large quantity of randomly distributed hydrogen held in the DLC structure and the associated van der Waals forces which have a low binding energy [88].

5.1.2. Soft (Hydrogenated a-C:H) and (Non-Hydrogenated a-C) Amorphous Carbon

These amorphous carbon films have properties similar to that of graphite, with most (exceeding 90 at. %) carbon atoms bound in sp^2 sites. The hardness of these DLCs can be less than 10 GPa. Such graphitic amorphous carbon coatings can be obtained through a variety of routes. These include carbon evaporation, low plasma density sputtering, plasma-assisted chemical vapour deposition (PACVD) at low plasma density or by the thermal decomposition of other forms of amorphous carbon [18, 89].

5.1.3. Hard (Hydrogenated a-C:H) Amorphous Carbon and Tetrahedral (ta-C) Carbon

When hydrogenated amorphous carbon (a-C:H) has a hydrogen content lower than 20 at. % combined with a high sp^3/sp^2 ratio the hardness of the material can approach 50 GPa [88,

90]. Being produced by plasma-enhanced chemical vapour deposition (PECVD) using hydrogen precursors, the manufactured films have a low friction coefficient.

Tetrahedral amorphous carbon (ta-C) films are mostly composed of sp^3 carbon sites, the proportion varying from 50% to 100%. These forms of DLC demonstrate the best tribological and mechanical properties of any DLC films. Consequently, the hardness in ta-C is the highest detected of any DLC; usually in the region of between 70 and 100 GPa which approaches that of diamond. Since ta-C is deposited as an amorphous carbon film, the surface is far smoother than coatings formed from polycrystalline diamond coatings. The interatomic bond energies are similar to those of bulk diamond, being higher than those in other DLC films. This explains the material's high hardness and atomic packing density and the relative ease to which they can be doped [18].

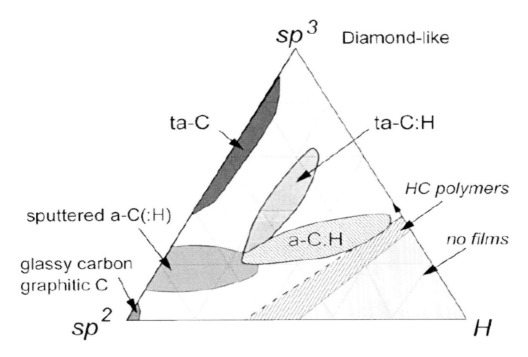

Figure 9. Phase diagram of diamond-like carbon materials. Reproduced with Permission From [91].

It can therefore be seen that both the concentration of hydrogen and the ratio of sp^3/sp^2 carbon atoms within the specific form of DLC reflects the properties of the material, with those forms having a higher $sp^3:sp^2$ ratio (e.g. ta-C) demonstrating more diamond-like character. In contrast, DLC films which are mostly composed of sp^2 carbon atoms (e.g. hydrogenated a:C-H) display more graphite-like character [92].

5.2. DLC Coating Methods

Numerous coating methods have been used for the growth of DLC films. These include cathodic arc spray, pulsed laser deposition, argon ion sputtering, ion beam deposition and plasma-enhanced chemical vapour deposition (PECVD).

The production of DLC films by ion sputtering in a vacuum has been shown to promote a more concentrated sp^3 carbon network, enhancing the diamond-like properties. The technique is also suitable for the coating of large surface areas [93]. However, the use of a vacuum can be prohibitive because of the expensive set-up costs involved and restrictions upon the shape or size of an object which can be treated. Therefore, various methods have been explored to perform DLC coating at atmospheric pressures to reduce the cost of producing effective coatings. One such technique involves electrochemical deposition using chemicals such as methanol and acetylene dissolved in ammonia. Naturally, this would involve the substrate being submerged in liquid which may not be preferable due to higher processing times associated with drying the finished products [94, 95].

However, PECVD is capable of producing a-C:H DLC films using a downstream acetylene, hydrogen and helium plasma. DLC produced via this method possessed an sp^3 carbon content of 57% [93]. What is noticeable is that not only was the procedure carried out at atmospheric pressure but also the processing temperature required for optimum results was only 200°C. This compares with diamond-coating temperatures of between 700 and 1000°C, illustrating the DLC deposited in this manner could be used as a coating on a far greater range of substrates [18].

Further improvements have seen it possible to deposit tetrahedral amorphous carbon films at room temperature using either infrared ($\lambda = 1064$ nm) [96], or ultraviolet ($\lambda = 248$ nm) pulsed laser deposition [97], ion beam deposition [98] or magnetron sputtering [99]. The properties and structure of films manufactured by high ion energy deposition techniques (UV-PLD and ion beam deposition) can be controlled by the kinetic energy of the carbon species can achieve this [18]. Furthermore, it has been shown in ion beam deposition that the incident angle of the kinetic carbon species plays an important role in the properties and structure of the resultant DLC film. If the ion beam makes contact with the surface at an acute angle, the actual ion energy present on the developing film significantly decreases, favouring the development of films with a higher sp^3 carbon content. Unfortunately this presents difficulties with the coating of three-dimensional products, where the incident angle of kinetic species will vary across the different faces of the specimen. Furthermore elevating the discharge voltage in ion beam deposited DLC films has been shown to increase the nanohardness whilst reducing the elastic modulus, in line with increased sp^3 content.

The deposition of DLC films by low ion energy deposition methods (IR-PLD) at room temperature in a reactive gas flow reactor was shown to be influenced by the reactive gas flow concentration (acetylene, C_2H_2 or nitrogen, N_2). The presence of nitrogen in the gas flow caused graphitisation to occur, whereas high concentrations of C_2H_2 favoured the production of a-C:H films; a more desirable outcome [100-103].

These low temperature reaction routes for the deposition of DLC have been cited as a more amenable method for the chemical stabilisation or functionalisation-preparation of delicate materials for applications such as biosensing and microelectromechanical systems (MEMS). This topic which will be covered in greater detail later in this review [37].

5.3. Properties of DLC

The desirable properties associated with DLC films have ensured its popularity as a biological coating. DLC possesses exceptional hardness, chemical inertness and high dielectric strength. Additional important properties also include its biocompatibility, both *in vitro* and *in vivo*, low surface roughness, infrared transparency and high electrical resistivity and refractive index [92, 93, 104, 105].

However, arguably the most useful properties of DLC which have been widely exploited are its low friction coefficient and excellent wear and abrasion resistance [15]. These characteristics can be explained by the production of a friction-induced interfacial transfer layer when the film is stressed. Formed in the uppermost region of a DLC coating, this layer possesses a much lower shear strength than the remaining DLC and exhibits the low friction and high wear resistance for which DLC is renowned. It is thought that friction-induced annealing, caused by the strain and thermal energy generated during sliding, is responsible for the transfer layer's existence [106, 107].

5.4. DLC Applications and its Use as a Biocoating

DLC enjoys widespread and varied applications throughout industry, even finding its way into everyday objects to enhance their usability and properties. However, it was not until the mid to late 1990s that industrial applications for the use of DLC truly took off. At first used in magnetic storage media, films were also applied to laser barcode scanners and eyeglasses to improve abrasion resistance [108]. In 1998 Gillette developed Mach 3 razor blades, which were manufactured with a DLC coating to improve the blade's quality and performance [15]. Additionally, Nissan has developed engines whose internal parts have a hydrogen-free DLC coating, on top of which sits a 2 nm layer of specially produced oil. This unique combination serves to reduce friction inside the moving parts of the engine by 25%, thus improving fuel efficiency and wear-resistance of the engine and lowering carbon emissions [109]. In addition some door hinges and lock mechanisms in cars are being coated with a DLC film rather than a thick layer of grease to aid lubrication [110].

DLC has also been utilised as an internal coating in polyethylene terephthalate (PET) bottles used in the drinks industry [111, 112]. The plastic has been used as an alternative to glass and metal beverage containers since 1976 due it being lightweight, less fragile, transparent and resealable. Unfortunately, however, PET is slightly permeable to gases, which transpires to any contents slowly oxidising as time progresses. At the same time, carbon dioxide can escape, leading to any carbonated drinks losing their fizz. These problems can be remedied through the application of a 50 nm DLC coating to the interior of any PET bottle preventing gaseous exchange. The technology, developed in Japan by Kirin in association with Mitsubishi Heavy Industries, is capable of coating up to 18,000 PET containers every hour, equating to 5 bottles per second. Such a rapid coating procedure is made possible through atmospheric plasma deposition. A colourless coating is deposited by increasing the band gap of the DLC to above 2.5 eV from the conventional value of 1.5 eV, which is aesthetically more desirable for the drinks industry. It was found that oxygen permeability could be reduced by a 23-fold magnitude in 1.5 litre bottles [113]. As such, these coatings are currently applied to PET bottles in Japan containing beverages with a relatively short shelf-life, such as tea, increasing their longevity.

DLC has also been applied as a biocoating in a variety of circumstances as illustrated by the following examples.

5.4.1. Orthodontic Wires

Orthodontic wires are used in dentistry for the correction of dental abnormalities, and are held in place by brackets. They are primarily used to close gaps between a patient's adjacent teeth and are normally manufactured from stainless steel or nickel-titanium alloys [11]. It has been suggested that a decreased level of friction between the bracket and wire might reduce the treatment period and further improve their anchorage control [114]. The use of plasma immersion ion implantation techniques has been used before in an effort to improve the mechanical properties and friction characteristics of orthodontic wires [115], but clinical trials have shown the technique offers no discernible difference over untreated wires [116]. In contrast, DLC coated (film thickness: 0.5 µm) orthodontic wires and brackets have been shown to reduce frictional resistance compared to those with untreated surfaces [11]. In a separate study, conducted by Kobayashi *et al.* [117], DLC was also shown to protect against the breakdown of nickel in corrosive environments, which could prove useful for patients who suffer from nickel allergies. Nanoindentation tests also confirmed that DLC coated wires have a lower elastic modulus than as-received wires, providing greater flexibility; a desirable characteristic for this application [11].

5.4.2. Vascular Stents

Patients suffering from arterial lesions or occlusions are recommended, in severe cases, to undergo endovascular surgical treatment. This minimally invasive technique involves the widening of the affected vessel, with the placement of a stent in the affected region. Nitinol (nickel-titanium alloy) stents having properties such as high elasticity, biocompatibility, high radial strength, 'shape-memory' and reduced foreshortening. These stens have become the preferential clinical option for the treatment of arterial occlusive disease [118-120].

However, in-stent restenosis caused by neointimal hyperplasia, remains a problem [119]. Within 6 months of the operation the incidence rate of restenosis varies between 20 – 40% for patients fitted with nitinol stents, with researchers stating that stents should be made more biocompatible to prevent this. Enhanced proliferative tissue response may be due to a contact allergy with nickel or the release of metallic ions from the stents. Subsequent *in vivo* studies of DLC-coated nitinol stents implanted into canine iliac arteries demonstrated a significant reduction in neointimal hyperplasia compared to the untreated stent [121]. In contrast, however, when polyethylene-glycol (PEG) was grafted on to the DLC coated stent surface, this caused a significant increase in hyperplasia, due to increased fibroblast proliferation, compared to the control and DLC coated stent surface alone. This contrasts with a previous *in vitro* study which concluded that the addition of PEG enhanced both uncoated, and DLC coated, nitinol stents [10]. These results illustrate that the use of DLC, however, is a useful addition when applied as a coating for stents, contributing to the device's biocompatibility.

5.4.3. Artificial Joint Implants

Patients requiring a total hip replacement are usually fitted with an artificial joint constructed from either metallic alloys or ceramics. The acetabular cup of the implant meanwhile is manufactured from ultra high molecular weight polyethylene (UHMWPE) [22]. Its use for conventional joint replacements has been widespread due to its chemical properties, price and biocompatibility [122].

However, UHMWPE does not exhibit long-term *in vivo* wear resistance, limiting the longevity of hip arthroplasties [122]. On average the UHMWPE component wears at an approximate rate of 0.1mm per year, ejecting 100 million wear particles into the joint space every day [123]. These particulates have been shown to initiate a cascade of adverse tissue responses ultimately cumulating to osteolysis and possible component-loosening [123]. Alternative polymers, such as highly crosslinked polyethylene (HXLPE) have also been suggested. Although the volume of wear particles would be greatly reduced if a hip implant was produced from such a material, the size of the debris would also be smaller, which may have the potential to stimulate biological activity and lead to increased osteolytic reactions. Research has therefore focused on the application of a corrosion and wear resistant layer, such as DLC, to artificial implants to prolong their life [22, 122]. Depositing a 700 nm hydrogenated DLC film by PECVD was found to be effective in significantly increasing the wear resistance of UHMWPE [122]. Similarly, if the metallic femoral head is coated with DLC, the wear rate of uncoated UHMWPE was also found to much lower. This can be attributed to the lower friction experienced between DLC and UHMWPE, compared to a metallic-UHMWPE interface. The fact that DLC prevents the metallic surface undergoing oxidation, further reducing the friction coefficient, is also noteworthy [22].

On a separate note, DLC coatings, especially when doped with phosphorous or silicon [124, 125], can improve the haemocompatability of implanted devices. This is thought to occur through the reduced incidence of inflammatory cytokines which would normally accompany the release of metal ions or wear particles [126]. This haemocompatibility could also be explained by the decrease in contact angle measurement between normal DLC (approximately 63°) compared to 5 at.% phosphorous doped DLC (P:DLC) (approximately 17°) . The lower the contact angle is, the lower the interfacial surface tension which, it has been suggested, could induce fewer conformational changes in any adsorbed plasma proteins [124].

5.4.4. Biosensors and Bioarrays

The chemical functionalisation of diamond and diamond-like carbon, which has been discussed previously (see section 2), yields some interesting opportunities with respect to the creation of biosensors and bioarrays. These applications require materials that are robust and stable, and DLC has been shown to be most suitable, surpassing the usability of alternative materials for these purposes, such as glass, silicon and gold. In fact, the signal to noise ratio is 40% higher on oligonucleotide arrays prepared on DLC compared to gold. This is due to the lower background fluorescence of amorphous carbon and a greater number of hybridisation sites [127].

Numerous alkene-containing molecules have been grafted via UV irradiation to amorphous carbon substrates to provide a number of different chemical surface functionalities including 1-dodecene, trifluroacetic acid protected 10-aminodec-1-ene (TFAAD), 10-N-Boc-amido-dec-1-ene and 9-decene-1-ol [34-37, 71, 127-130]. Of these, TFAAD has seen the most widespread usage in research for the manufacture of biosensors upon carbon substrates. Once the TFAAD molecule has been deprotected it produces a free amine group onto which biomolecules can be attached (see Figure 10). Amine groups are preferable for bioconjugation since they do not readily oxidise, negating the requirement for any immediate preparation and purification or unnecessary wastage [127].

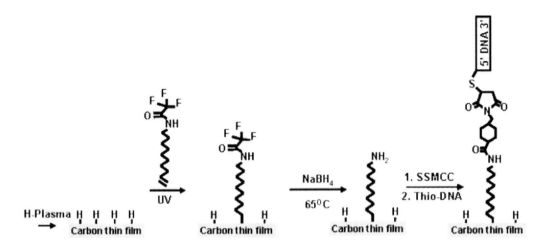

Figure 10. Functionalisation of amorphous carbon to produce a TFAAD-modified surface, subsequent deprotection to yield an amine-modified surface which is reacted with an SSMCC linker and thio-DNA to produce a covalently bonded DNA adduct on the amorphous carbon surface. Reproduced in altered form with permission from [37].

It has been shown that DLC substrates can withstand the conditions necessary for in situ oligonucleotide synthesis. They are also able to better withstand high temperatures, basic conditions or repeated hybridisation cycles compared to arrays produced on glass [131]. It has been further demonstrated that oligonucleotide arrays produced on DLC are just as stable as those manufactured on diamond thin films [37, 127].

5.4.4. Cell Growth Patterning

Cell types including fibroblasts, osteoblasts, retinal pericytes, endothelial cells and glial-like cell lines have been successfully cultured on DLC *in vitro* [8, 9, 124, 132-134]. Emerging research has illustrated that the characteristics of DLC may be altered to either favour or deter the attachment and proliferation of certain cell types. This alteration, usually though the doping of elements, demonstrates the varying degree to which different cells can be bound to its surface. Kelly *et al.* [104] illustrated that rat primary cortical neurones had poor adhesion to pure DLC, instead preferring to form free-floating neurospheres. In contrast, the doping of DLC with between 5 and 20% phosphorous (P:DLC) causes the surface to be highly hydrophilic, encouraging the adsorption of poly-L-lysine, a typical coating for neuronal cultures [92]. Consequently, neurones cultured on P:DLC form a coherent monolayer [104]. The production of a patterned grid of P:DLC upon a DLC substrate favoured the growth of the cells solely along the phosphorus-doped lines as seen in Figure 11 [92].

It has been mentioned that the doping of phosphorous on DLC improves its haemocompatibility, possibly due to a reduction in the surface contact angle. Experimentation has shown that DLC possesses a contact angle of between 60 - 70°, whilst that of P:DLC has been recorded between 16 - 33°, the value depending on to production process and the doping level [104, 124]. The decreased contact angle produces a film with a lower interfacial surface tension, which leads to any absorbed plasma proteins undergoing fewer conformational changes and less denaturing [135]. With respect to neuronal cell cultures, poly-L-lysine has been observed to have greater adsorption on P:DLC and O:DLC compared to DLC, enhancing neurone attachment and growth [104].

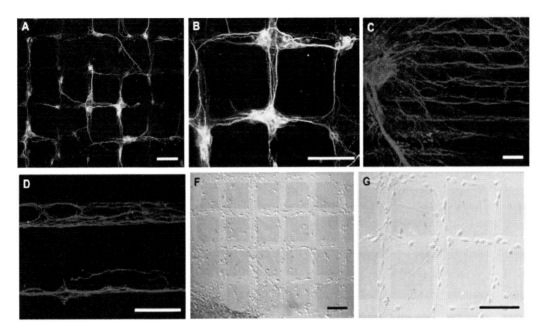

Figure 11. Preferential patterning of cortical neurons (A,B: Green=MAP2 Stain, Blue=DAPI), dorsal root ganglion cells (C,D: Red = β-III tubulin, Blue=DAPI) and human neural progenitors (E,F) cultured along P:DLC tracks. Scale bar: 90 μm. Reproduced with permission from [92].

Microcontact printing has also proved effective in directing the growth of neuronal cells on diamond and DLC [134]. Polydimethylsiloxane stamps were coated with protein solutions of laminin, an extracellular matrix protein which enhances the attachment and proliferation of neurons. These proteins were applied to diamond surfaces upon which murine cortical neurons were cultured for up to 8 days. Neuronal outgrowth occurred primarily along the laminin lines deposited by the stamps with cells failing to attach to any other untreated areas of the diamond surface. No degradation of the pattern occurred over the length of the experiment illustrating that the stamping technique is a relatively simple yet effective method of controlling directional cell growth.

Further work has analysed the possibility that patterned DLC surfaces could stimulate the differentiation of stem cells [136]. Surface topographical cues were found to encourage the production of neuronal-like cells from human bone marrow derived mesenchymal stem cells (hBM-MSCs) *in vitro* [136]. The deposition of DLC ridges of width 40 μm enhance the production of Brain-derived neurotrophic factor (BDNF), which in turn stimulated the production of nitric oxide (NO) which modulates neuronal function and controls neural differentiation. These mechanically stimulated elongated cells displayed long neurite-like protrusions and increased levels of NO comparable to differentiating neuronal cells. They also tested positive for the neuronal cell marker TUJ1 through immunofluorescence. Similarly, nanocrystalline diamond films, mentioned in further detail in section 3 exert a similar neuronal differentiating effect upon stem cells [54, 64].

To conclude, over the past couple of decades, the use and application of DLC in biological research has continued to expand. Its recognition as a non-toxic material, which can be deposited at room temperature to a wide variety of substrates to confer its desirable tribological properties, has seen use in numerous situations. The ability to graft functional groups and molecules to its surface is of great importance for the production of biosensors, a field which will undoubtedly become of much greater importance in the years to come as technology advances.

6. CARBON NANOPARTICLES

Carbon nanodots (C-dots), as shown in Figure 12 by high resolution TEM, form a recently discovered group of nanocarbons which are composed of quasispherical nanoparticles of diameters less than 10 nm [137]. They were first discovered, accidentally, through the purification of single-walled carbon nanotubes (SWCNTs) and have attracted widespread research appeal due to their intriguing properties [138]. Importantly, C-dots are naturally carboxylated when produced by arc-discharge methods, enabling them to be water soluble and also capable of undergoing functionalisation with numerous organic/biological species. Their well-defined morphology, nanoscale dimensions, ability for chemical functionalisation and a variety of inexpensive and simple production routes provides a promising platform from which C-dots can hope to compete with other carbon forms to contribute to future biological research. Of notable interest is the possible replacement of quantum dots (QDs) with C-dots. QDs are metal based nanometre sized particles which are frequently used as fluorescent probes for *in vitro* imaging. However, QDs are often produced from inherently toxic compounds (e.g. Cd and Se) which make their translation to *in vivo* and clinical use rather problematic. Given this disadvantage, important research has been directed into more biocompatible alternatives such as nanodiamond and carbon nanoparticles.

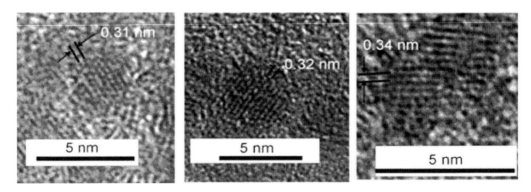

Figure 12. High resolution transmission electron microscope (HRTEM) images of C-Dots. Reproduced with permission from [144].

Synthesis of C-dots can occur through either top-down or bottom-up production methods [137]. In top-down procedures, such as arc discharge [138], laser ablation [139] and electrochemical oxidation [140], the C-dots are formed from the disintegration of a larger carbon structure. Meanwhile, bottom-up methods, consisting of thermal [141], microwave

[142] or supported synthetic [141, 143] approaches, produce C-dots from molecular precursors. Once C-dots have been isolated they are normally oxidised using nitric acid and then purified further through the use of a separation technique, namely centrifugation or electrophoresis [137]. Further details on the production methods of C-dots are detailed a recent review by Baker and Baker [137].

In contrast to nanodiamond particles, C-dots exhibit more graphite-like character, due to the greater abundance of sp^2 hybridised carbon atoms in their structure. They have a relatively high oxygen content, earning them the alternative name of carbogenic nanodots [137]. It is interesting to note that the production method used to synthesise C-dots can determine the particle's physical and chemical properties. Manufacture of C-dots through thermal reactions of citrate salts produces relatively large particles (approximately 7 nm), with X-ray diffraction (XRD) analysis illustrating a disordered carbon structure [141], whereas C-dots synthesised by the electrochemical oxidation of MWCNTs were largely graphitic, having sp^2 and disordered carbon structures [144].

6.1. Applications of Carbon Nanoparticles

One of the main research areas of C-dots focuses on understanding and taking advantage of their innate photoluminescence abilities. C-dots generally absorb radiation in the UV spectrum, possessing excitation edges at between 270-280 nm [144, 145]. Moreover, the precise emission wavelength is known to depend upon the intensity and wavelength of the excitation source. The exact reasons for this phenomenon remain to be answered. It is speculated that it could be the result of quantum confinement, whereby differently sized nanoparticles are optically selected or possibly due to the different emissive 'traps' on the surface of the C-dots [137]. However, it is these photoluminescence properties which make C-dots such an interesting proposition for future biological applications, as mentioned below:

6.1.1. In Vitro Bioimaging

C-dots have become an attractive alternative to biologically hazardous quantum dots such as cadmium selenide (CdSe) for bioimaging purposes. Dependent upon the fabrication method used to produce the C-dots, surface passivation may be required to produce luminescence [137]. This is true in the case of laser ablation synthesised C-dots, where it is speculated that surface energy traps present on the C-dots emit photon energy following passivation. This passivation occurs through the adsorption of an organic moiety such as 4,7,10-trioxa-1,13-tridecanediamine (TTDDA) or poly-ethylene glycol (PEG) [146, 147] which serves to increase C-dot absorbance/fluorescence in between 350 – 550 nm.

Initial biological experiments on the use of C-dots focused on the imagining of cancer cells. Sun *et al.* [139] utilised C-dots which had been passivated using poly(propionyl-ethylenimine-*co*-ethylenimine) PPEI-EI to image human breast cancer MCF-7 cells using two-photon luminescence microscopy. After incubation with the C-dots in media for 2 hours at 37°C, and subsequent washing, it was discovered the cell membrane and cytoplasm exhibited vivid green luminescence, although the C-dots failed to migrate into the nuclei. It is assumed that the particles were taken up by the cells through endocytosis, although at cooler temperatures (4°C) uptake of the nanodots did not occur.

Cellular imaging using C-dots produced from candle soot has been also been achieved [148] using Ehrlich ascites carcinoma cells (EACs) (see Figure 13). The C-dots were produced from the chemically homogenised soot of a burnt candle, producing 2 – 6 nm particles. These C-dots were incubated with the cells for 30 minutes, after which the remaining C-dots were separated from the solution by centrifugation, and subsequently imaged without the need for surface passivation to occur. The imaging of cells as diverse as E. Coli bacteria and murine progenitor cells has been achieved in a similar fashion [143]. All these samples exhibited fluorescence which was photostable and did not readily undergo photobleaching. It is further anticipated in the future that C-dots could pass through the nuclear membrane of cells through their attachment to facilitator proteins or peptides, allowing for the effect of C-dots on nuclei to be effectively researched [137].

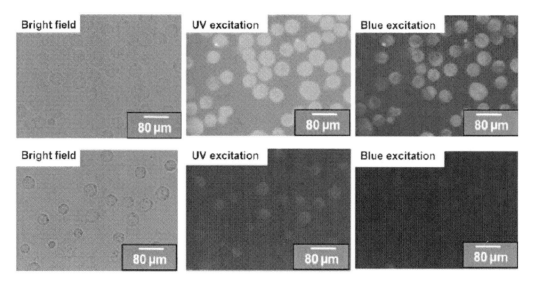

Figure 13. C-Dot labelling of Ehrlich's Ascites Carcinoma (EAC) cells. The bottom row depicts control images where no C-dots were used. Cells were washed and observed under bright field, UV and blue light excitations. Cells became a bright turquoise colour under UV excitation and yellow under blue excitation (exhibiting no fluorescence in the control). The cells appear light blue under UV excitation due to cellular autofluorescence. Reproduced with permission from [148].

6.1.2. In Vivo Biolabelling

In order for a biological label to be suitable for *in vivo* use, it should non-toxic, biocompatible, photostable and sufficiently intense so to be easily detected. Murine models have been procured to test the *in vivo* efficacy and efficiency of C-dots and zinc sulphide (ZnS) doped C-dots passivated with polyethylene glycol (PEG) [149]. The particles were injected subcutaneously, intravenously and intradermally into female DBA/1 mice (see Figure 14) and could be excited through light of wavelengths 470 nm and 545 nm.

Following intradermal delivery, both forms of C-dots were found to migrate to the auxillary lymph nodes, although the doped C-dots travelled at a slower rate, possibly as a result of PEG functionalisation restricting interaction with the murine lymph cells. After intravenous application the C-dots were seen to migrate around the body of the mice via the bloodstream. Photoluminescence was detected from the bladder with C-dot emission from the urine being recorded 3 hours after injection. After four hours the organs were harvested, with the C-dots having accumulated in the kidney, with significantly fewer present in the liver of the mice. This is to be expected following the standard urine pathway, however it was expected that greater numbers of nanoparticles would be located in the liver, in accordance with previous studies on *in vivo* nanoparticle behaviour [151]. However, this discrepancy may have been the result of the PEG functionalisation, which decreased the protein affinity of the C-dots, causing them to pass relatively quickly to the kidney for excretion.

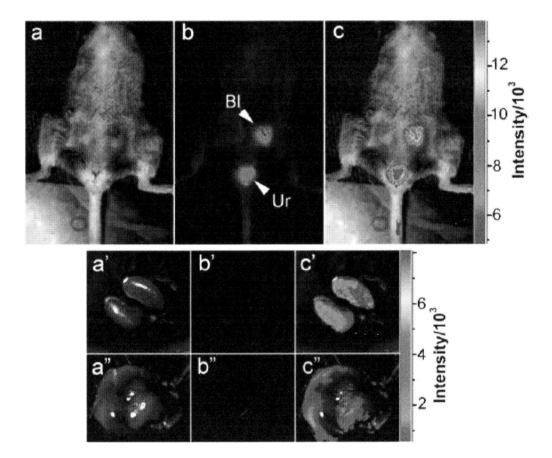

Figure 14. Intravenous injection of C-Dots: (A) bright field, (B) as-detected fluorescence (Bl = bladder; Ur = urine), and (C) colour-coded images. The same order is used for the images of the dissected kidneys (A' – C') and the liver (A'' – C''). Reproduced with permission from [150].

It is therefore possible to conclude that C-dots hold a great deal of promise for biological research. The wide availability of inexpensive renewable resources for their production (lignocellulosic biomass wastes being one example) [137] and a number of synthetic manufacturing methods enables them to be produced with relative ease. The advantages attributed to C-dots including their biocompatibility, colloidal stability, chemical stability and photobleaching resistance illustrates their potential in future optical imaging applications. As research in the field continues to progress and mature the importance and influence attributed to C-dots will undoubtedly increase.

CONCLUSION

In this review we have seen that diamond and its related materials are particularly well suited to be used in biological research. The ability of such materials to be deposited either as thin films (DLC, (U)NCD) enables them to be utilised as surface coatings, either in their pure, original form (DLC and ND coatings on artificial prostheses to enhance biocompatibility and reduce wear) or functionalised using a range of different chemical and biological moieties to give rise to biosensors and MEMs.

The existence of diamond materials in a nanoparticulate form allows for even greater applicability, particularly in the field of bioimaging where fluorescent NDs and C-dots hold particular advantages (lack of cytotoxicity and photobleaching) over existing conventional imaging materials. Furthermore the possible use of diamond and carbon nanoparticles as drug and gene carriers can hold particular significance for the future of drug administration, allowing for less invasive, more effective therapies.

In conclusion, the different forms of diamond we have discussed each possess unique advantages which are currently opening the door to a range of intriguing possibilities in biomedical research. In the future it is anticipated that the role which diamond has secured as a biomaterial will only increase in importance as the range of applications for which it is suited become more widely accepted and recognised.

REFERENCES

[1] Shimkunas, R. A.; Robinson, E.; Lam, R.; Lu, S.; Xu, X. Y.; Zhang, X. Q.; Huang, H. J.; Osawa, E.; Ho, D., Nanodiamond-insulin complexes as pH-dependent protein delivery vehicles. *Biomaterials* 2009, *30* (29), 5720-5728.

[2] Chow, E. K.; Zhang, X.-Q.; Chen, M.; Lam, R.; Robinson, E.; Huang, H.; Schaffer, D.; Osawa, E.; Goga, A.; Ho, D., Nanodiamond Therapeutic Delivery Agents Mediate Enhanced Chemoresistant Tumor Treatment. *Science Translational Medicine* 2011, *3* (73), 73ra21.

[3] Mostofizadeh, A.; Li, Y. W.; Song, B.; Huang, Y. D., Synthesis, Properties, and Applications of Low-Dimensional Carbon-Related Nanomaterials. *Journal of Nanomaterials* 2011.

[4] Narayan, R. J.; Boehm, R. D.; Sumant, A. V., Medical applications of diamond particles & surfaces. *Materials Today* 2011, *14* (4), 154-163.

[5] Zhang, Q. W.; Mochalin, V. N.; Neitzel, I.; Knoke, I. Y.; Han, J. J.; Klug, C. A.; Zhou, J. G.; Lelkes, P. I.; Gogotsi, Y., Fluorescent PLLA-nanodiamond composites for bone tissue engineering. *Biomaterials* 2011, *32* (1), 87-94.

[6] Xiao, X. C.; Wang, J.; Liu, C.; Carlisle, J. A.; Mech, B.; Greenberg, R.; Guven, D.; Freda, R.; Humayun, M. S.; Weiland, J.; Auciello, O., In vitro and in vivo evaluation of ultrananocrystalline diamond for coating of implantable retinal microchips. *J. Biomed. Mater. Res. Part B* 2006, *77B* (2), 273-281.

[7] Thalhammer, A.; Edgington, R. J.; Cingolani, L. A.; Schoepfer, R.; Jackman, R. B., The use of nanodiamond monolayer coatings to promote the formation of functional neuronal networks. *Biomaterials* 2010, *31* (8), 2097-2104.

[8] Allen, M.; Myer, B.; Rushton, N., In vitro and in vivo investigations into the biocompatibility of diamond-like carbon (DLC) coatings for orthopedic applications. *Journal of Biomedical Materials Research* 2001, *58* (3), 319-328.

[9] Singh, A.; Ehteshami, G.; Massia, S.; He, J. P.; Storer, R. G.; Raupp, G., Glial cell and fibroblast cytotoxicity study on plasma-deposited diamond-like carbon coatings. *Biomaterials* 2003, *24* (28), 5083-5089.

[10] Shin, H. S.; Park, K.; Kim, J. H.; Kim, J. J.; Han, D. K.; Moon, M. W.; Lee, K. R.; Shin, J. H., Biocompatible PEG Grafting on DLC-coated Nitinol Alloy for Vascular Stents. *Journal of Bioactive and Compatible Polymers* 2009, *24* (4), 316-328.

[11] Muguruma, T.; Iijima, M.; Brantley, W. A.; Mizoguchi, I., Effects of a diamond-like carbon coating on the frictional properties of orthodontic wires. *Angle Orthodontist* 2011, *81* (1), 141-148.

[12] Martin, R.; Alvaro, M.; Herance, J. R.; Garcia, H., Fenton-Treated Functionalized Diamond Nanoparticles as Gene Delivery System. *Acs Nano* 2010, *4* (1), 65-74.

[13] Martin, R.; Menchon, C.; Apostolova, N.; Victor, V. M.; Alvaro, M.; Herance, J. R.; Garcia, H., Nano-Jewels in Biology. Gold and Platinum on Diamond Nanoparticles as Antioxidant Systems Against Cellular Oxidative Stress. *Acs Nano* 2010, *4* (11), 6957-6965.

[14] Ho, D. A., Beyond the Sparkle: The Impact of Nanodiamonds as Biolabeling and Therapeutic Agents. *Acs Nano* 2009, *3* (12), 3825-3829.

[15] Grill, A., Diamond-like carbon: state of the art. *Diam. Relat. Mat.* 1999, *8* (2-5), 428-434.

[16] Liu, K. K.; Cheng, C. L.; Chang, C. C.; Chao, J. I., Biocompatible and detectable carboxylated nanodiamond on human cell. *Nanotechnology* 2007, *18* (32).

[17] Dubrovinskaia, N.; Dubrovinsky, L.; Crichton, W.; Langenhorst, F.; Richter, A., Aggregated diamond nanorods, the densest and least compressible form of carbon. *Applied Physics Letters* 2005, *87* (8).

[18] Neuville, S.; Matthews, A., A perspective on the optimisation of hard carbon and related coatings for engineering applications. *Thin Solid Films* 2007, *515* (17), 6619-6653.

[19] Yang, W. S.; Auciello, O.; Butler, J. E.; Cai, W.; Carlisle, J. A.; Gerbi, J.; Gruen, D. M.; Knickerbocker, T.; Lasseter, T. L.; Russell, J. N.; Smith, L. M.; Hamers, R. J., DNA-modified nanocrystalline diamond thin-films as stable, biologically active substrates. *Nature Materials* 2002, *1* (4), 253-257.

[20] Ganesan, K.; Stacey, A.; Meffin, H.; Lichter, S.; Greferath, U.; Fletcher, E. L.; Prawer, S., Diamond penetrating electrode array for Epi-Retinal Prosthesis. *Conf Proc IEEE Eng Med Biol Soc* 2010, *1*, 6757-60.

[21] Jones, B. J.; Mahendran, A.; Anson, A. W.; Reynolds, A. J.; Bulpett, R.; Franks, J., Diamond-like carbon coating of alternative metal alloys for medical and surgical applications. *Diam. Relat. Mat.* 2010, *19* (7-9), 685-689.

[22] Grill, A., Diamond-like carbon coatings as biocompatible materials - an overview. *Diam. Relat. Mat.* 2003, *12* (2), 166-170.

[23] Tzeng, Y. K.; Faklaris, O.; Chang, B. M.; Kuo, Y.; Hsu, J. H.; Chang, H. C., Superresolution Imaging of Albumin-Conjugated Fluorescent Nanodiamonds in Cells by Stimulated Emission Depletion. *Angewandte Chemie-International Edition* 2011, *50* (10), 2262-2265.

[24] Kong, X. L.; Huang, L. C. L.; Hsu, C. M.; Chen, W. H.; Han, C. C.; Chang, H. C., High-affinity capture of proteins by diamond nanoparticles for mass spectrometric analysis. *Analytical Chemistry* 2005, *77* (1), 259-265.

[25] Huang, L. C. L.; Chang, H. C., Adsorption and immobilization of cytochrome c on nanodiamonds. *Langmuir* 2004, *20* (14), 5879-5884.

[26] Huang, H. J.; Pierstorff, E.; Osawa, E.; Ho, D., Protein-mediated assembly of nanodiamond hydrogels into a biocompatible and biofunctional multilayer nanofilm. *Acs Nano* 2008, *2* (2), 203-212.

[27] Steinmuller-Nethl, D.; Kloss, F. R.; Najam-U-Haq, M.; Rainer, M.; Larsson, K.; Linsmeier, C.; Koehler, G.; Fehrer, C.; Lepperdinger, G.; Liu, X.; Memmel, N.; Bertel, E.; Huck, C. W.; Gassner, R.; Bonn, G., Strong binding of bioactive BMP-2 to nanocrystalline diamond by physisorption. *Biomaterials* 2006, *27* (26), 4547-4556.

[28] Bondar, V. S.; Pozdnyakova, I. O.; Puzyr, A. P., Applications of nanodiamonds for separation and purification of proteins. *Physics of the Solid State* 2004, *46* (4), 758-760.

[29] Chung, P. H.; Perevedentseva, E.; Tu, J. S.; Chang, C. C.; Cheng, C. L., Spectroscopic study of bio-functionalized nanodiamonds. *Diam. Relat. Mat.* 2006, *15* (4-8), 622-625.

[30] Huang, T. S.; Tzeng, Y.; Liu, Y. K.; Chen, Y. K.; Walker, K. R.; Guntupalli, R.; Liu, C., Immobilization of antibodies and bacterial binding on nanodiamond and carbon nanotubes for biosensor applications. *Diam. Relat. Mat.* 2004, *13* (4-8), 1098-1102.

[31] Kong, X. L.; Huang, L. C. L.; Liau, S. C. V.; Han, C. C.; Chang, H. C., Polylysine-coated diamond nanocrystals for MALDI-TOF mass analysis of DNA oligonucleotides. *Analytical Chemistry* 2005, *77* (13), 4273-4277.

[32] Wang, H. D.; Niu, C. H.; Yang, Q. Q.; Badea, I., Study on protein conformation and adsorption behaviors in nanodiamond particle-protein complexes. *Nanotechnology* 2011, *22* (14).

[33] Kruger, A.; Liang, Y. J.; Jarre, G.; Stegk, J., Surface functionalisation of detonation diamond suitable for biological applications. *Journal of Materials Chemistry* 2006, *16* (24), 2322-2328.

[34] Krueger, A., New carbon materials: Biological applications of functionalized nanodiamond materials. *Chemistry-a European Journal* 2008, *14* (5), 1382-1390.

[35] Nichols, B. M.; Metz, K. M.; Tse, K. Y.; Butler, J. E.; Russell, J. N.; Hamers, R. J., Electrical bias dependent photochemical functionalization of diamond surfaces. *Journal of Physical Chemistry B* 2006, *110* (33), 16535-16543.

[36] Nichols, B. M.; Butler, J. E.; Russell, J. N.; Hamers, R. J., Photochemical functionalization of hydrogen-terminated diamond surfaces: A structural and mechanistic study. *Journal of Physical Chemistry B* 2005, *109* (44), 20938-20947.

[37] Sun, B.; Colavita, P. E.; Kim, H.; Lockett, M.; Marcus, M. S.; Smith, L. M.; Hamers, R. J., Covalent photochemical functionalization of amorphous carbon thin films for integrated real-time biosensing. *Langmuir* 2006, *22* (23), 9598-9605.

[38] Lud, S. Q.; Steenackers, M.; Jordan, R.; Bruno, P.; Gruen, D. M.; Feulner, P.; Garrido, J. A.; Stutzmann, M., Chemical grafting of biphenyl self-assembled monolayers on ultrananocrystalline diamond. *Journal of the American Chemical Society* 2006, *128* (51), 16884-16891.

[39] Baker, S. E.; Tse, K. Y.; Hindin, E.; Nichols, B. M.; Clare, T. L.; Hamers, R. J., Covalent functionalization for biomolecular recognition on vertically aligned carbon nanofibers. *Chemistry of Materials* 2005, *17* (20), 4971-4978.

[40] Kuo, T. C.; McCreery, R. L.; Swain, G. M., Electrochemical modification of boron-doped chemical vapor deposited diamond surfaces with covalently bonded monolayers. *Electrochemical and Solid State Letters* 1999, *2* (6), 288-290.

[41] Brooksby, P. A.; Downard, A. J., Multilayer nitroazobenzene films covalently attached to carbon. An AFM and electrochemical study. *Journal of Physical Chemistry B* 2005, *109* (18), 8791-8798.

[42] Allongue, P.; Delamar, M.; Desbat, B.; Fagebaume, O.; Hitmi, R.; Pinson, J.; Saveant, J. M., Covalent modification of carbon surfaces by aryl radicals generated from the electrochemical reduction of diazonium salts. *Journal of the American Chemical Society* 1997, *119* (1), 201-207.

[43] Spitsyn, B. V.; Davidson, J. L.; Gradoboev, M. N.; Galushko, T. B.; Serebryakova, N. V.; Karpukhina, T. A.; Kulakova, II; Melnik, N. N., Inroad to modification of detonation nanodiamond. *Diam. Relat. Mat.* 2006, *15* (2-3), 296-299.

[44] Sotowa, K. I.; Amamoto, T.; Sobana, A.; Kusakabe, K.; Imato, T., Effect of treatment temperature on the amination of chlorinated diamond. *Diam. Relat. Mat.* 2004, *13* (1), 145-150.

[45] Liu, Y.; Gu, Z. N.; Margrave, J. L.; Khabashesku, V. N., Functionalization of nanoscale diamond powder: Fluoro-, alkyl-, amino-, and amino acid-nanodiamond derivatives. *Chemistry of Materials* 2004, *16* (20), 3924-3930.

[46] Krueger, A.; Stegk, J.; Liang, Y.; Lu, L.; Jarre, G., Biotinylated Nanodiamond: Simple and Efficient Functionalization of Detonation Diamond. *Langmuir* 2008, *24* (8), 4200-4204.

[47] Meinhardt, T.; Lang, D.; Dill, H.; Krueger, A., Pushing the Functionality of Diamond Nanoparticles to New Horizons: Orthogonally Functionalized Nanodiamond Using Click Chemistry. *Advanced Functional Materials* 2011, *21* (3), 494-500.

[48] Yeap, W. S.; Chen, S. M.; Loh, K. P., Detonation Nanodiamond: An Organic Platform for the Suzuki Coupling of Organic Molecules. *Langmuir* 2009, *25* (1), 185-191.

[49] Osswald, S.; Yushin, G.; Mochalin, V.; Kucheyev, S. O.; Gogotsi, Y., Control of sp(2)/sp(3) carbon ratio and surface chemistry of nanodiamond powders by selective oxidation in air. *Journal of the American Chemical Society* 2006, *128* (35), 11635-11642.

[50] Tsugawa, K.; Ishihara, M.; Kim, J.; Koga, Y.; Hasegawa, M., Nanocrystalline diamond film growth on plastic substrates at temperatures below 100 degrees C from low-temperature plasma. *Physical Review B* 2010, *82* (12).

[51] Tiwari, R. N.; Chang, L., Growth, microstructure, and field-emission properties of synthesized diamond film on adamantane-coated silicon substrate by microwave plasma chemical vapor deposition. *Journal of Applied Physics* 2010, *107* (10).

[52] Ihara, M.; Maeno, H.; Miyamoto, K.; Komiyama, H., DIAMOND DEPOSITION ON SILICON SURFACES HEATED TO TEMPERATURE AS LOW AS 135-DEGREES-C. *Applied Physics Letters* 1991, *59* (12), 1473-1475.

[53] Gruen, D. M., NANOCRYSTALLINE DIAMOND FILMS1. *Annual Review of Materials Science* 1999, *29* (1), 211-259.

[54] Chen, Y. C.; Lee, D. C.; Tsai, T. Y.; Hsiao, C. Y.; Liu, J. W.; Kao, C. Y.; Lin, H. K.; Chen, H. C.; Palathinkal, T. J.; Pong, W. F.; Tai, N. H.; Lin, I. N.; Chiu, I. M., Induction and regulation of differentiation in neural stem cells on ultra-nanocrystalline diamond films. *Biomaterials* 2010, *31* (21), 5575-5587.

[55] Williams, O. A.; Daenen, M.; D'Haen, J.; Haenen, K.; Maes, J.; Moshchalkov, V. V.; Nesladek, M.; Gruen, D. M., Comparison of the growth and properties of ultrananocrystalline diamond and nanocrystalline diamond. *Diam. Relat. Mat.* 2006, *15* (4-8), 654-658.

[56] Chong, K. F.; Loh, K. P.; Vedula, S. R. K.; Lim, C. T.; Sternschulte, H.; Steinmüller, D.; Sheu, F.-s.; Zhong, Y. L., Cell Adhesion Properties on Photochemically Functionalized Diamond. *Langmuir* 2007, *23* (10), 5615-5621.

[57] Eckert, M.; Neyts, E.; Bogaerts, A., Differences between Ultrananocrystalline and Nanocrystalline Diamond Growth: Theoretical Investigation of CxHy Species at Diamond Step Edges. *Crystal Growth & Design* 2010, *10* (9), 4123-4134.

[58] Chen, Y. C.; Zhong, X. Y.; Konicek, A. R.; Grierson, D. S.; Tai, N. H.; Lin, I. N.; Kabius, B.; Hiller, J. M.; Sumant, A. V.; Carpick, R. W.; Auciello, O., Synthesis and characterization of smooth ultrananocrystalline diamond films via low pressure bias-enhanced nucleation and growth. *Applied Physics Letters* 2008, *92* (13).

[59] Balda, M. S.; Matter, K., Epithelial cell adhesion and the regulation of gene expression. *Trends in Cell Biology* 2003, *13* (6), 310-318.

[60] Zutter, M. M., Integrin-mediated adhesion: Tipping the balance between chemosensitivity and chemoresistance. *Breast Cancer Chemosensitivity* 2007, *608*, 87-100.

[61] Khang, D.; Lu, J.; Yao, C.; Haberstroh, K. M.; Webster, T. J., The role of nanometer and sub-micron surface features on vascular and bone cell adhesion on titanium. *Biomaterials* 2008, *29* (8), 970-983.

[62] Armstrong, R. J. E.; Svendsen, C. N., Neural stem cells: From cell biology to cell replacement. *Cell Transplantation* 2000, *9* (2), 139-152.

[63] Bithell, A.; Williams, B. P., Neural stem cells and cell replacement therapy: making the right cells. *Clinical Science* 2005, *108* (1), 13-22.

[64] Chen, Y. C.; Lee, D. C.; Hsiao, C. Y.; Chung, Y. F.; Chen, H. C.; Thomas, J. P.; Pong, W. F.; Tai, N. H.; Lin, I. N.; Chiu, I. M., The effect of ultra-nanocrystalline diamond films on the proliferation and differentiation of neural stem cells. *Biomaterials* 2009, *30* (20), 3428-3435.

[65] Lechleitner, T.; Klauser, F.; Seppi, T.; Lechner, J.; Jennings, P.; Perco, P.; Mayer, B.; Steinmuller-Nethl, D.; Preiner, J.; Hinterdorfer, P.; Hermann, M.; Bertel, E.; Pfaller, K.; Pfaller, W., The surface properties of nanocrystalline diamond and nanoparticulate diamond powder and their suitability as cell growth support surfaces. *Biomaterials* 2008, *29* (32), 4275-4284.

[66] Clem, W. C.; Chowdhury, S.; Catledge, S. A.; Weimer, J. J.; Shaikh, F. M.; Hennessy, K. M.; Konovalov, V. V.; Hill, M. R.; Waterfeld, A.; Bellis, S. L.; Vohra, Y. K., Mesenchymal stem cell interaction with ultra-smooth nanostructured diamond for wear-resistant orthopaedic implants. *Biomaterials* 2008, *29* (24-25), 3461-3468.

[67] Shi, B.; Jin, Q.; Chen, L.; Auciello, O., Fundamentals of ultrananocrystalline diamond (UNCD) thin films as biomaterials for developmental biology: Embryonic fibroblasts growth on the surface of (UNCD) films. *Diam. Relat. Mat.* 2009, *18* (2-3), 596-600.

[68] Amaral, M.; Dias, A. G.; Gomes, P. S.; Lopes, M. A.; Silva, R. F.; Santos, J. D.; Fernandes, M. H., Nanocrystalline diamond: In vitro biocompatibility assessment by MG63 and human bone marrow cells cultures. *Journal of Biomedical Materials Research Part A* 2008, *87A* (1), 91-99.

[69] Huang, H.; Chen, M.; Bruno, P.; Lam, R.; Robinson, E.; Gruen, D.; Ho, D., Ultrananocrystalline Diamond Thin Films Functionalized with Therapeutically Active Collagen Networks. *The Journal of Physical Chemistry B* 2009, *113* (10), 2966-2971.

[70] Hartl, A.; Schmich, E.; Garrido, J. A.; Hernando, J.; Catharino, S. C. R.; Walter, S.; Feulner, P.; Kromka, A.; Steinmuller, D.; Stutzmann, M., Protein-modified nanocrystalline diamond thin films for biosensor applications. *Nat Mater* 2004, *3* (10), 736-742.

[71] Stavis, C.; Clare, T. L.; Butler, J. E.; Radadia, A. D.; Carr, R.; Zeng, H.; King, W. P.; Carlisle, J. A.; Aksimentiev, A.; Bashir, R.; Hamers, R. J., Surface functionalization of thin-film diamond for highly stable and selective biological interfaces. *Proceedings of the National Academy of Sciences* 2011, *108* (3), 983-988.

[72] Radadia, A. D.; Stavis, C. J.; Carr, R.; Zeng, H.; King, W. P.; Carlisle, J. A.; Aksimentiev, A.; Hamers, R. J.; Bashir, R., Control of Nanoscale Environment to Improve Stability of Immobilized Proteins on Diamond Surfaces. *Advanced Functional Materials* 2011, *21* (6), 1040-1050.

[73] Abouzar, M. H.; Poghossian, A.; Razavi, A.; Besmehn, A.; Bijnens, N.; Williams, O. A.; Haenen, K.; Wagner, P.; Schoning, M. J., Penicillin detection with nanocrystalline-diamond field-effect sensor. *Physica Status Solidi a-Applications and Materials Science* 2008, *205* (9), 2141-2145.

[74] Bergonzo, P.; Bongrain, A.; Scorsone, E.; Bendali, A.; Rousseau, L.; Lissorgues, G.; Mailley, P.; Li, Y.; Kauffmann, T.; Goy, F.; Yvert, B.; Sahel, J. A.; Picaud, S., 3D shaped mechanically flexible diamond microelectrode arrays for eye implant applications: The MEDINAS project. *IRBM* 2011, *32* (2), 91-94.

[75] Chan, H. Y.; Aslam, D. M.; Wiler, J. A.; Casey, B., A Novel Diamond Microprobe for Neuro-Chemical and -Electrical Recording in Neural Prosthesis. *Journal of Microelectromechanical Systems* 2009, *18* (3), 511-521.

[76] Xing, Y.; Xiong, W.; Zhu, L.; Osawa, E.; Hussin, S.; Dai, L. M., DNA Damage in Embryonic Stem Cells Caused by Nanodiamonds. *Acs Nano* 2011, *5* (3), 2376-2384.

[77] Zhu, L.; Chang, D. W.; Dai, L. M.; Hong, Y. L., DNA damage induced by multiwalled carbon nanotubes in mouse embryonic stem cells. *Nano Letters* 2007, *7* (12), 3592-3597.

[78] Liu, K. K.; Wang, C. C.; Cheng, C. L.; Chao, J. I., Endocytic carboxylated nanodiamond for the labeling and tracking of cell division and differentiation in cancer and stem cells. *Biomaterials* 2009, *30* (26), 4249-4259.

[79] Schrand, A. M.; Huang, H. J.; Carlson, C.; Schlager, J. J.; Osawa, E.; Hussain, S. M.; Dai, L. M., Are diamond nanoparticles cytotoxic? *Journal of Physical Chemistry B* 2007, *111* (1), 2-7.

[80] Vista Vista Engineering: Diamond Research. http://www.vistaeng.com/?page_id=81 (accessed 2011).

[81] Dolmatov, V. Y., Detonation synthesis ultradispersed diamonds: Properties and applications. *Uspekhi Khimii* 2001, *70* (7), 687-708.

[82] Davidson, J. L.; Kang, W. P., Applying CVD diamond and particulate nanodiamond. *Synthesis, Properties and Applications of Ultrananocrystalline Diamond* 2005, *192*, 357-372.

[83] Red'kin, V. E., Lubricants with ultradisperse diamond-graphite powder. *Chemistry and Technology of Fuels and Oils* 2004, *40* (3), 164-170.

[84] Fu, C. C.; Lee, H. Y.; Chen, K.; Lim, T. S.; Wu, H. Y.; Lin, P. K.; Wei, P. K.; Tsao, P. H.; Chang, H. C.; Fann, W., Characterization and application of single fluorescent nanodiamonds as cellular biomarkers. *Proceedings of the National Academy of Sciences of the United States of America* 2007, *104* (3), 727-732.

[85] Yu, S. J.; Kang, M. W.; Chang, H. C.; Chen, K. M.; Yu, Y. C., Bright fluorescent nanodiamonds: No photobleaching and low cytotoxicity. *Journal of the American Chemical Society* 2005, *127* (50), 17604-17605.

[86] Huang, H.; Pierstorff, E.; Osawa, E.; Ho, D., Active nanodiamond hydrogels for chemotherapeutic delivery. *Nano Letters* 2007, *7*, 3305-3314.

[87] Nair, V., Retrovirus-induced oncogenesis and safety of retroviral vectors. *Current Opinion in Molecular Therapeutics* 2008, *10* (5), 431-438.

[88] Weissmantel, C.; Bewilogua, K.; Breuer, K.; Dietrich, D.; Ebersbach, U.; Erler, H. J.; Rau, B.; Reisse, G., Preparation And Properties Of Hard I-C And I-BN Coatings. *Thin Solid Films* 1982, *96* (1), 31-44.

[89] Koidl, P., Wild, C.,Lacher,R.,Sah,R.F.,, *Diamond and Diamond Like Films and Coatings*. Plenum Press: New York, USA, 1991.

[90] Tamor, M. A.; Vassell, W. C.; Carduner, K. R., Atomic Constraint In Hydrogenated Diamond-Like Carbon. *Applied Physics Letters* 1991, *58* (6), 592-594.

[91] Robertson, J., Diamond-like amorphous carbon. *Materials Science and Engineering: R: Reports* 2002, *37* (4-6), 129-281.

[92] Regan, E. M.; Uney, J. B.; Dick, A. D.; Zhang, Y. W.; Nunez-Yanez, J.; McGeehan, J. P.; Claeyssens, F.; Kelly, S., Differential patterning of neuronal, glial and neural progenitor cells on phosphorus-doped and UV irradiated diamond-like carbon. *Biomaterials* 2010, *31* (2), 207-215.

[93] Ladwig, A. M.; Koch, R. D.; Wenski, E. G.; Hicks, R. F., Atmospheric plasma deposition of diamond-like carbon coatings. *Diam. Relat. Mat.* 2009, *18* (9), 1129-1133.

[94] Izake, E. L.; Paulmier, T.; Bell, J. M.; Fredericks, P. M., Characterization of reaction products and mechanisms in atmospheric pressure plasma deposition of carbon films from ethanol. *Journal of Materials Chemistry* 2005, *15* (2), 300-306.

[95] Novikov, V. P.; Dymont, V. P., Synthesis of diamondlike films by an electrochemical method at atmospheric pressure and low temperature. *Applied Physics Letters* 1997, *70* (2), 200-202.

[96] Cappelli, E.; Scilletta, C.; Mattei, G.; Valentini, V.; Orlando, S.; Servidori, M., Critical role of laser wavelength on carbon films grown by PLD of graphite. *Applied Physics A: Materials Science & Processing* 2008, *93* (3), 751-758.

[97] Orlianges, J. C.; Champeaux, C.; Catherinot, A.; Pothier, A.; Blondy, P.; Abelard, P.; Angleraud, B., Electrical properties of pure and metal doped pulsed laser deposited carbon films. *Thin Solid Films* 2004, *453-454*, 291-295.

[98] Ronning, C.; Buttner, M.; Vetter, U.; Feldermann, H.; Wondratschek, O.; Hofsass, H.; Brunner, W.; Au, F. C. K.; Li, Q.; Lee, S. T., Ion beam deposition of fluorinated amorphous carbon. *Journal of Applied Physics* 2001, *90* (8), 4237-4245.

[99] Hastas, N. A.; Dimitriadis, C. A.; Logothetidis, S.; Angelis, C. T.; Konofaos, N.; Evangelou, E. K., Temperature dependence of the barrier at the tetrahedral amorphous carbon-silicon interface. *Semiconductor Science and Technology* 2001, *16* (6), 474-477.

[100] Kahn, M.; Cekada, M.; Berghauser, R.; Waldhauser, W.; Bauer, C.; Mitterer, C.; Brandstatter, E., Accurate Raman spectroscopy of diamond-like carbon films deposited by an anode layer source. *Diam. Relat. Mat.* 2008, *17* (7-10), 1647-1651.

[101] Kahn, M.; Cekada, M.; Schoberl, T.; Berghauser, R.; Mitterer, C.; Bauer, C.; Waldhauser, W.; Brandstatter, E., Structural and mechanical properties of diamond-like carbon films deposited by an anode layer source. *Thin Solid Films* 2009, *517* (24), 6502-6507.

[102] Kahn, M.; Menegazzo, N.; Mizaikoff, B.; Berghauser, R.; Lackner, J. M.; Hufnagel, D.; Waldhauser, W., Properties of DLC and Nitrogen-Doped DLC Films Deposited by DC Magnetron Sputtering. *Plasma Processes and Polymers* 2007, *4*, S200-S204.

[103] Kahn, M.; Paskvale, S.; Cekada, M.; Schoberl, T.; Waldhauser, W.; Mitterer, C.; Pelicon, P.; Brandstatter, E., The relationship between structure and mechanical properties of hydrogenated amorphous carbon films. *Diam. Relat. Mat.* 2010, *19* (10), 1245-1248.

[104] Kelly, S.; Regan, E. M.; Uney, J. B.; Dick, A. D.; McGeehan, J. P.; Mayer, E. J.; Claeyssens, F.; Bristol Biochip, G., Patterned growth of neuronal cells on modified diamond-like carbon substrates. *Biomaterials* 2008, *29* (17), 2573-2580.

[105] Wang, H. Y.; Xu, M.; Zhang, W.; Kwok, D. T. K.; Jiang, J. A.; Wu, Z. W.; Chu, P. K., Mechanical and biological characteristics of diamond-like carbon coated poly aryl-ether-ether-ketone. *Biomaterials* 2010, *31* (32), 8181-8187.

[106] Grill, A., Tribology of diamondlike carbon and related materials: an updated review. *Surface & Coatings Technology* 1997, *94-5* (1-3), 507-513.

[107] Erdemir, A.; Bindal, C.; Pagan, J.; Wilbur, P., Characterization of transfer layers on steel surfaces sliding against diamond-like hydrocarbon films in dry nitrogen. *Surface & Coatings Technology* 1995, *76-77* (1-3), 559-563.

[108] Petrmichl, R. Diamond-Like Carbon as a Protective Coating for Decorative Glass *Glass Performance Days 2009 Conference* [Online], 2009, p. 613. http://www.glassglobal.com/gpd/downloads/Coatings-Petrmichl.pdf.

[109] Nissan Ultra-Low Friction Diamond Like Carbon. http://www.nissan-global.com/EN/Technology/Overview/dlc.html (accessed 13 April).

[110] Erdemir, A.; Donnet, C., Tribology of diamond-like carbon films: recent progress and future prospects. *J. Phys. D-Appl. Phys.* 2006, *39* (18), R311-R327.

[111] Boutroy, N.; Pernel, Y.; Rius, J. M.; Auger, F.; Bardeleben, H. J. v.; Cantin, J. L.; Abel, F.; Zeinert, A.; Casiraghi, C.; Ferrari, A. C.; Robertson, J., Hydrogenated amorphous carbon film coating of PET bottles for gas diffusion barriers. *Diam. Relat. Mat.* 2006, *15* (4-8), 921-927.

[112] Shirakura, A.; Nakaya, M.; Koga, Y.; Kodama, H.; Hasebe, T.; Suzuki, T., Diamond-like carbon films for PET bottles and medical applications. *Thin Solid Films* 2006, *494* (1-2), 84-91.

[113] Ueda, A.; Nakachi, M.,Goto, S.,Yamakoshi, H.,Shirakura, A.,, High Speed and High Gas Barrier Rotary DLC Plasma Coating System for PET Bottles. *Mitsubishi Heavy Industries Ltd Technical Review* 2005, *42* (1), 1-3.

[114] Burrow, S. J., Friction and resistance to sliding in orthodontics: A critical review. *American Journal of Orthodontics and Dentofacial Orthopedics* 2009, *135* (4), 442-447.

[115] Wichelhaus, A.; Geserick, M.; Hibst, R.; Sander, F. G., The effect of surface treatment and clinical use on friction in NiTi orthodontic wires. *Dental Materials* 2005, *21* (10), 938-945.

[116] Cobb, N. W.; Kula, K. S.; Phillips, C.; Proffit, W. R., Efficiency of multi-strand steel, superelastic Ni-Ti and ion-implanted Ni-Ti archwires for initial alignment. *Clin Orthod Res* 1998, *1* (1), 12-9.

[117] Kobayashi, S.; Ohgoe, Y.; Ozeki, K.; Hirakuri, K.; Aoki, H., Dissolution effect and cytotoxicity of diamond-like carbon coatings on orthodontic archwires. *Journal of Materials Science-Materials in Medicine* 2007, *18*, 2263-2268.

[118] Schillinger, M.; Sabeti, S.; Loewe, C.; Dick, P.; Amighi, J.; Mlekusch, W.; Schlager, O.; Cejna, M.; Lammer, J.; Minar, E., Calloon angioplasty versus implantation of nitinol stents in the superficial femoral artery. *New England Journal of Medicine* 2006, *354* (18), 1879-1888.

[119] Duda, S. H.; Bosiers, M.; Lammer, J.; Scheinert, D.; Zeller, T.; Tielbeek, A.; Anderson, J.; Wiesinger, B.; Tepe, G.; Lansky, A.; Mudde, C.; Tielemans, H.; Beregi, J. P., Sirolimus-eluting versus bare nitinol Stent for obstructive superficial femoral artery disease: The SIROCCO II trial. *Journal of Vascular and Interventional Radiology* 2005, *16* (3), 331-338.

[120] Ponec, D.; Jaff, M. R.; Swischuk, J.; Feiring, A.; Laird, J.; Mehra, M.; Popma, J. J.; Investigators, C. S., The nitinol SMART stent vs Wallstent for suboptimal iliac artery angioplasty: CRISP-US trial results. *Journal of Vascular and Interventional Radiology* 2004, *15* (9), 911-918.

[121] Kim, J. H.; Shin, J. H.; Shin, D. H.; Moon, M. W.; Park, K.; Kim, T. H.; Shin, K. M.; Won, Y. H.; Han, D. K.; Lee, K. R., Comparison of diamond-like carbon-coated nitinol stents with or without polyethylene glycol grafting and uncoated nitinol stents in a canine iliac artery model. *British Journal of Radiology* 2011, *84* (999), 210-215.

[122] Puertolas, J. A.; Martinez-Nogues, V.; Martinez-Morlanes, M. J.; Mariscal, M. D.; Medel, F. J.; Lopez-Santos, C.; Yubero, F., Improved wear performance of ultra high molecular weight polyethylene coated with hydrogenated diamond like carbon. *Wear* 2010, *269* (5-6), 458-465.

[123] Sinha, R. K., *Hip Replacement: Current Trends and Controversies*. Marcel Dekker, Inc: New York, USA, 2005.

[124] Kwok, S. C. H.; Jin, W.; Chu, P. K., Surface energy, wettability, and blood compatibility phosphorus doped diamond-like carbon films. *Diam. Relat. Mat.* 2005, *14* (1), 78-85.

[125] Ong, S. E.; Zhang, S.; Du, H.; Too, H. C.; Aung, K. N., Influence of silicon concentration on the haemocompatibility of amorphous carbon. *Biomaterials* 2007, *28* (28), 4033-4038.

[126] Ma, W. J.; Ruys, A. J.; Mason, R. S.; Martin, P. J.; Bendavid, A.; Liu, Z. W.; Ionescu, M.; Zreiqat, H., DLC coatings: Effects of physical and chemical properties on biological response. *Biomaterials* 2007, *28* (9), 1620-1628.

[127] Lockett, M. R.; Shortreed, M. R.; Smith, L. M., Aldehyde-terminated amorphous carbon substrates for the fabrication of biomolecule arrays. *Langmuir* 2008, *24* (17), 9198-9203.

[128] Colavita, P. E.; Sun, B.; Wang, X. Y.; Hamers, R. J., Influence of Surface Termination and Electronic Structure on the Photochemical Grafting of Alkenes to Carbon Surfaces. *Journal of Physical Chemistry C* 2009, *113* (4), 1526-1535.

[129] Hamers, R. J.; Coulter, S. K.; Ellison, M. D.; Hovis, J. S.; Padowitz, D. F.; Schwartz, M. P.; Greenlief, C. M.; Russell, J. N., Cycloaddition chemistry of organic molecules with semiconductor surfaces. *Accounts Chem. Res.* 2000, *33* (9), 617-624.

[130] Strother, T.; Knickerbocker, T.; Russell, J. N.; Butler, J. E.; Smith, L. M.; Hamers, R. J., Photochemical functionalization of diamond films. *Langmuir* 2002, *18* (4), 968-971.

[131] Phillips, M. F.; Lockett, M. R.; Rodesch, M. J.; Shortreed, M. R.; Cerrina, F.; Smith, L. M., In situ oligonucleotide synthesis on carbon materials: stable substrates for microarray fabrication. *Nucleic Acids Research* 2008, *36* (1).

[132] Okpalugo, T. I. T.; McKenna, E.; Magee, A. C.; McLaughlin, J.; Brown, N. M. D., The MTT assays of bovine retinal pericytes and human microvascular endothelial cells on DLC and Si-DLC-coated TCPS. *Journal of Biomedical Materials Research Part A* 2004, *71A* (2), 201-208.

[133] Kumari, T. V.; Kumar, P. R. A.; Muraleedharan, C. V.; Bhuvaneshwar, G. S.; Sampeur, Y.; Derangere, F.; Suryanarayanan, R., In vitro cytocompatibility studies of Diamond Like Carbon coatings on titanium. *Bio-Medical Materials and Engineering* 2002, *12* (4), 329-338.

[134] Specht, C. G.; Williams, O. A.; Jackman, R. B.; Schoepfer, R., Ordered growth of neurons on diamond. *Biomaterials* 2004, *25* (18), 4073-4078.

[135] Kwok, S. C. H.; Ha, P. C. T.; McKenzie, D. R.; Bilek, M. M. M.; Chu, P. K., Biocompatibility of calcium and phosphorus doped diamond-like carbon thin films synthesized by plasma immersion ion implantation and deposition. *Diam. Relat. Mat.* 2006, *15* (4-8), 893-897.

[136] D'Angelo, F.; Armentano, I.; Mattioli, S.; Crispoltoni, L.; Tiribuzi, R.; Cerulli, G. G.; Palmerini, C. A.; Kenny, J. M.; Martino, S.; Orlacchio, A., Micropatterned Hydrogenated Amorphous Carbon Guides Mesenchymal Stem Cells Towards Neuronal Differentiation. *European Cells & Materials* 2010, *20*, 231-244.

[137] Baker, S. N.; Baker, G. A., Luminescent Carbon Nanodots: Emergent Nanolights. *Angewandte Chemie-International Edition* 2010, *49* (38), 6726-6744.

[138] Xu, X. Y.; Ray, R.; Gu, Y. L.; Ploehn, H. J.; Gearheart, L.; Raker, K.; Scrivens, W. A., Electrophoretic analysis and purification of fluorescent single-walled carbon nanotube fragments. *Journal of the American Chemical Society* 2004, *126* (40), 12736-12737.

[139] Cao, L.; Wang, X.; Meziani, M. J.; Lu, F. S.; Wang, H. F.; Luo, P. J. G.; Lin, Y.; Harruff, B. A.; Veca, L. M.; Murray, D.; Xie, S. Y.; Sun, Y. P., Carbon dots for multiphoton bioimaging. *Journal of the American Chemical Society* 2007, *129* (37), 11318-+.

[140] Lu, J.; Yang, J. X.; Wang, J. Z.; Lim, A. L.; Wang, S.; Loh, K. P., One-Pot Synthesis of Fluorescent Carbon Nanoribbons, Nanoparticles, and Graphene by the Exfoliation of Graphite in Ionic Liquids. *Acs Nano* 2009, *3* (8), 2367-2375.

[141] Bourlinos, A. B.; Stassinopoulos, A.; Anglos, D.; Zboril, R.; Karakassides, M.; Giannelis, E. P., Surface functionalized carbogenic quantum dots. *Small* 2008, *4* (4), 455-458.

[142] Zhu, H.; Wang, X.; Li, Y.; Wang, Z.; Yang, F.; Yang, X., Microwave synthesis of fluorescent carbon nanoparticles with electrochemiluminescence properties. *Chem Commun (Camb)* 2009, (34), 5118-20.

[143] Liu, R. L.; Wu, D. Q.; Liu, S. H.; Koynov, K.; Knoll, W.; Li, Q., An Aqueous Route to Multicolor Photoluminescent Carbon Dots Using Silica Spheres as Carriers. *Angewandte Chemie-International Edition* 2009, *48* (25), 4598-4601.

[144] Zhou, J. G.; Booker, C.; Li, R. Y.; Zhou, X. T.; Sham, T. K.; Sun, X. L.; Ding, Z. F., An electrochemical avenue to blue luminescent nanocrystals from multiwalled carbon nanotubes (MWCNTs). *Journal of the American Chemical Society* 2007, *129*, 744-745.

[145] Hu, S. L.; Niu, K. Y.; Sun, J.; Yang, J.; Zhao, N. Q.; Du, X. W., One-step synthesis of fluorescent carbon nanoparticles by laser irradiation. *Journal of Materials Chemistry* 2009, *19* (4), 484-488.

[146] Peng, H.; Travas-Sejdic, J., Simple Aqueous Solution Route to Luminescent Carbogenic Dots from Carbohydrates. *Chemistry of Materials* 2009, *21* (23), 5563-5565.

[147] Qiao, Z. A.; Wang, Y. F.; Gao, Y.; Li, H. W.; Dai, T. Y.; Liu, Y. L.; Huo, Q. S., Commercially activated carbon as the source for producing multicolor photoluminescent carbon dots by chemical oxidation. *Chemical Communications* 2010, *46* (46), 8812-8814.

[148] Ray, S. C.; Saha, A.; Jana, N. R.; Sarkar, R., Fluorescent Carbon Nanoparticles: Synthesis, Characterization, and Bioimaging Application. *Journal of Physical Chemistry C* 2009, *113* (43), 18546-18551.

[149] Yang, S. T.; Wang, X.; Wang, H. F.; Lu, F. S.; Luo, P. J. G.; Cao, L.; Meziani, M. J.; Liu, J. H.; Liu, Y. F.; Chen, M.; Huang, Y. P.; Sun, Y. P., Carbon Dots as Nontoxic and High-Performance Fluorescence Imaging Agents. *Journal of Physical Chemistry C* 2009, *113* (42), 18110-18114.

[150] Yang, S. T.; Cao, L.; Luo, P. G. J.; Lu, F. S.; Wang, X.; Wang, H. F.; Meziani, M. J.; Liu, Y. F.; Qi, G.; Sun, Y. P., Carbon Dots for Optical Imaging in Vivo. *Journal of the American Chemical Society* 2009, *131* (32), 11308-+.

[151] Li, S. D., 7th International Symposium on Polymer Therapeutics, Valencia, Spain. *American Chemical Society* 2007, 496.

Chapter 10

PLASMA ENHANCED CVD DIAMOND COATINGS ON TRANSITION METAL SUBSTRATES: AN INTERFACIAL CHEMISTRY STUDY BY SYNCHROTRON RADIATION

Yuanshi Li and Akira Hirose*
Plasma Physics Lab, University of Saskatchewan, Saskatoon, SK, Canada

ABSTRACTS

Diamond coating on transition metal substrates including the typical Fe, Co, Ni-based alloys, and Cu, Ti alloys has been an important research topic because it can combine the advantages of diamond coatings, which are superhard, highly thermal conductive, chemically inert and wear resistant, with those of the underlying substrates including low materials cost, superior mechanical/physical performances and excellent machinability. However, it is difficult to obtain high quality and well adherent diamond coatings on these substrates because several key technical barriers. During the deposition process, complex interfacial reaction can occur between the gaseous precursors and the substrate components due to the high reactivity, solubility or diffusivity of carbon and/or hydrogen with the metal elements. The other problem is that the huge mismatch of the thermal expansion coefficients between the diamond coating and substrate materials that cause severe delamination of the coatings produced due to accumulated internal stress. We have performed comprehensive investigations on the diamond deposition on a series of transition metal and their alloys in terms of the nucleation, growth, and adhesion ability. Different deposition performances and growth mechanism are observed. In this chapter, we will introduce our new progress on the diamond nucleation and growth on the conventional transition metals and, especially, the interfacial chemistry and fine structures of the diamond coatings and the transition metal interfaces that have been addressed by TEM and synchrotron radiation.

---

* Corresponding author: Fax: 1-306-966-6400; E-mail: yul088@mail.usask.ca (Y. Li), Plasma Physics Lab, University of Saskatchewan, Saskatoon, SK, S7N 5E2, Canada

1. Introduction

1.1. Properties of Diamond

Diamond is one of the most promising materials for advanced industrial applications such as in heat sink, optical devices, electronic devices, wear resistant protective coatings due to its extraordinary properties involving mechanical, physical, chemical and many other advantages [1]. Diamond is the hardest known material and has the highest bulk modulus and the lowest compressibility. It has the lowest coefficient of thermal expansion at room temperature and a high thermal conductivity. In addition, diamond is electrically insulating and optically transparent from deep ultra-violet to far infrared. Diamond is resistant to chemical corrosion and biologically compatible [2]. Given these unique properties, diamond has been regarded as a 21st-centrury high performance material that has a good prospect of application and extension.

1.2. Current Situation of Diamond Coating on Hetero-Substrates

Single crystalline, polycrystalline, nanocrystalline diamond films have been artificially synthesized at sub-atmospheric pressures on various hetero-substrates such as metal, ceramics, polymer, etc, by chemical vapor deposition (CVD) techniques. This novel processing technique provides an economically viable alternative to the traditional high temperature high pressure method, and has significantly extended application areas of diamond as protective or functional coating materials. Three classes of hetero-substrate materials can be sub-divided regarding the basic carbon-substrate interaction:

a. Little or no carbon solubility or reaction. These materials include such metals as Cu, Sn, Pb, Ag, and Au as well as non-metals Ge, sapphire, and graphite.
b. Carbide Formers. These include metals such as Ti, Zr, Hf, Si, V, Nb, Ta, Cr, Mo, Al, Mn, W, Fe, Co, Ni, Y and some other rare earth metals.
c. Significant C diffusion. In this case, the substrate acts as a carbon sink and the deposited carbon dissolves into the metal substrate and forms a solid solution or carbide. Such metals include Pt, Pd, Fe, Co, Ni, Mn and Ti.

CVD of diamond coatings on non-diamond substrates usually involves an initial formation of interfacial layer of carbide upon which the diamond subsequently grows [3]. Therefore, it is difficult to synthesize fast-growing and well adherent diamond thin films directly on those first class metals including Cu, Au, Ag, because of the lack of carbide formation between diamond and the metal surface. The second class metals among the above mentioned materials are considered to be favorable for rapid production of diamond film. Particularly, Si has been selected for long as a standard material due to its thermal and chemical compatibility to diamond which ensures adequate nucleation density, growth rates and adhesion ability of diamond thin films. Some researchers have confirmed that diamond nucleates and grows on the top of a SiC intermediate layer formed during the initial stage of the deposition process. In addition, Si-containing compounds such as SiO_2, quartz, and Si_3N_4

also form carbide layers. For some metals like Ti and Cr, the TiC or CrxCy carbide layer grown at the early stage of diamond deposition can be several tens or even hundred μm thick [4,5]. Such a thick interfacial carbide layer might significantly affect the mechanical properties, and the utility of CVD diamond coatings on these materials must be very careful. Substrates composed of carbides themselves, such as SiC, WC, TiC and Si_3N_4-SiC are particularly amenable to diamond deposition. The third class metals Fe, Co, Ni and their alloys and compounds like WC-Co, carbon steel, stainless steels, and tool steels, stand for the least successful substrates for deposition of adherent diamond films, which demonstrate slow nucleation rate, low nucleation density, and poor adhesion strength of diamond films. This chapter has primarily focused on our most recent achievements regarding successful coating of high quality diamond films on such transition metal substrates, which has been once considered difficult or impossible.

2. DIAMOND COATINGS ON TRANSITION METALS

2.1. Advantages of Diamond Coatings

In comparison with other substrate materials, transition metals are the most commonly used and cost-effective structural materials in many aspects of modern industry, and they have been widely applied as taps, dies, twist drills, reamers, saw blades, cutting tools, thermal sinks. The properties of these materials can be greatly adjusted by designing the initial composition and optimizing the microstructures through subsequent thermo-mechanical treatment. Basically, when they are used as critical components in harsh (wear, corrosive and erosive) environments, accelerated damage easily occurs. As the early failure is usually initiated from the outer-most surface, it is important and cost-effective to obtain strengthened high performance surface for longevity of service.

It is a promising idea to deposit dense, continuous and well adherent diamond films on such bulk materials in order to obtain engineered surface properties, without deteriorating the strength and toughness of the substrates. The duplex diamond-metal or alloys system will yield properties much superior to those of individual substrate and diamond, allowing enhanced product performance and lifetime. For instance, the diamond coated substrate can possess increased surface hardness and anticorrosion behavior while retaining its tough core. Previous study indicated that a several microns thick diamond film could increase the surface hardness up to 80 GPa and the lifetime of 304 SS was even increased by factors of several hundreds against high impact wear [1]. A one micron diamond coating also increased the resistance of steel to low impact wear from abrasives by a factor of 60. This represents significant increases in the life of the steel substrates. In addition, the successful coating of high quality diamond thin film with controlled structures on steel substrates will provide cost-effective substitutes for cemented carbides and other hard tools for industrial applications, as well as prevent steels from humidity, abrasion, corrosion or other deformations in various industrial applications [6,7].

2.2. Barriers of Diamond Coating on Transition Metals

Even though the CVD synthesis of diamond films on a range of substrate materials like silicon wafers and ceramics, has become very successful, CVD diamond on transition metal substrates (titanium, chromium, iron, cobalt, nickel, copper) and their alloys is still a great challenge due to problems associated with nucleation, growth and adhesion [8,9]. The typical limiting factors making the direct nucleation, growth and good adhesion of diamond film on transition metals troublesome include:

a) High diffusion coefficient of carbon in Fe, Ni, Co matrix and high reactivity with Fe, Ti, Cr
   Metals such as tungsten, titanium, tantalum and zirconium react with carbon to produce carbides, while iron, cobalt, nickel and manganese dissolve carbon. As a result, under diamond growth conditions, carbon dissolves into the steel matrix to form a solid solution, leading to a temporary decrease in the surface carbon concentration. Titanium is also very reactive with hydrogen and the formation of brittle titanium hydride can greatly damage the surface mechanical properties of the substrate by inducing fatigue cracking. Consequently, the onset of diamond nucleation is delayed due to the lack of a critical carbon supersaturation required for the diamond nucleation.
b) Fe/Ni/Co acts as a catalyst for the formation of graphite.
   Furthermore, carbon is too reactive with the Fe/Ni/Co elements and under diamond deposition conditions, these metals act as catalysts for the preferential formation of graphitic phase, which continues to remain as soot at the interface between the diamond film and steel substrates, leading to low nucleation rate/density and poor adhesion of diamond film on the steel substrate. The typical images are illustrated in Figure 1 which clearly demonstrates the formation of a graphite intermediate layer and a deep internal penetration zone of carbon while diamond film is quite difficult to remain on the steel surface.
c) The huge mismatch in the thermal expansion coefficients between transition metals/their alloys and the diamond film.
   Another major constraint in preventing the formation of continuous and adherent diamond films on steel is the large difference in their thermal expansion coefficients. The high elastic modulus of diamond leads to high thermal stresses during the process of heating up and cooling down, resulting in an increased tendency for crinkle, cracking, curliness and/or delamination of the diamond film. The large difference in the thermal expansion coefficient of diamond and steel also leads to very poor adhesion. As a combined consequence of the above detrimental factors, direct deposition of uniform and adherent diamond films on ferrous-base substrates without any special pretreatment has been scarcely reported [10].

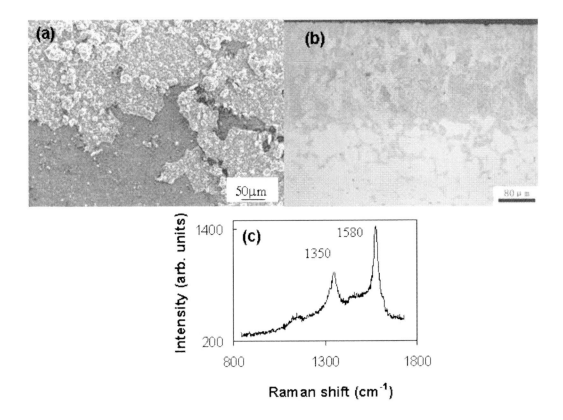

Figure 1. Typical surface (a) and cross sectional (b) SEM images of Fe-base alloy after CVD diamond deposition in CH4-H2 mixture. (c) Raman spectrum of the diamond-substrate interface.

3. TECHNICAL APPROACHES

3.1. Direct Coating of Adherent Diamond Films on Steels

A direct deposition of continuous and adherent diamond films on steel substrates has been extensively explored in our recent research project. The goal is likely to be achieved by modifying the steel using suitable alloying elements considering their different affinity with the reaction gases and diamond. The related work on the direct fabrication of diamond films on various bulk steel substrates of different compositions revealed that the nucleation, growth, and adhesion properties of the diamond film are strongly dependent on the types and relative concentrations of alloy elements in the steel substrates. As a direct consequence, we have realized successful fabrication of high quality diamond films on specific steel substrates without using any interlayer or pretreatment.

3.1.1. Effects of Types of Alloying Elements

Various alloying elements Si, Al, Cr, Ti, Mo and Mn, have been individually added into Fe-Cr ferritic steel. The CVD growth of diamond on these substrates was performed at 670 °C in a microwave plasma enhanced CVD reactor [11]. The results indicated that the diamond films grown on Fe-Cr substrates modified by Si, Cr, Mo, Ti, which have been successfully employed as metal substrates or interlayers for high quality diamond film growth, contain a high fraction of non-diamond phases and easily suffer a spontaneous detachment from the steel substrate after CVD process. Instead, Al is the only satisfactory alloying element to assist in direct coating of continuous and well adherent diamond films by successfully preventing the formation of graphic carbon between the diamond and steel interface. Such a beneficial effect of Al is shown in Figure 2. A thickness approximately 3 μm, has been achieved on this modified substrate (Fig. 3). The adhesion strength of such continuous diamond thin films has been evaluated using several different techniques including indentation test and scratch test (Fig. 4). Significantly increased surface hardness (over ten times higher than the bulk steel substrate) and enhanced wear resistance have been provided by these diamond films. The important conclusion drawn from this research is that Al is a promising alloying element to ensure an effective surface engineering of steels by CVD diamond coatings. Al also demonstrates great potential as high performance intermediate material for diamond coating on Al-free bulk steels, as discussed later.

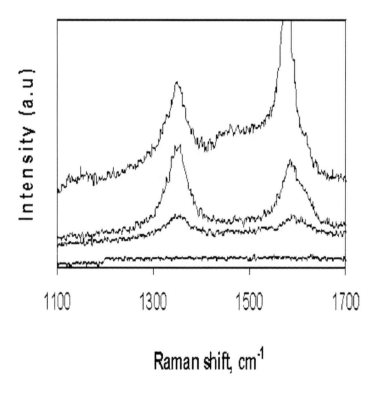

Figure 2. Raman spectra of diamond film-steel interface, showing intermediate graphic carbon is effectively suppressed with increasing Al concentrations in the substrate.

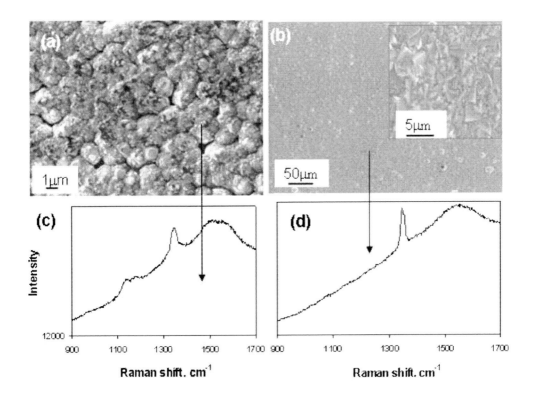

Figure 3. SEM images (a,b) and Raman spectra (c,d) of diamond films directly grown on Fe-Cr-Al steels for 4 (a, c) and 12 (b, d) hrs.

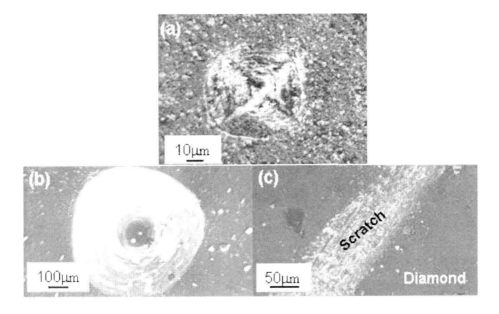

Figure 4. SEM images of diamond film deposited on Fe-Cr-Al steels; (a) with Vickers Microindentation test; (b) after Rockwell C indentation test; and (c) after scratch test.

3.1.2 Effects of Substrate Compositions

A continued research indicates that the diamond films grown on Fe-base alloys show distinctive features depending on the types and relative concentrations of the alloying elements in the substrates [12,13]. The following observations have been obtained:

1. Poor quality diamond films are produced on carbon steel, Fe-Cr, Fe-Cr-Ni steels. Well adherent diamond films are directly obtainable on Fe-Cr-Al type steels. A higher Al concentration in binary Fe-Al alloys is also beneficial to enhance the nucleation density, nucleation rate, and adhesion of diamond films.
2. A pre-oxidation treatment of Fe-Ni-Al alloy to form an Al_2O_3 barrier improves the diamond quality. A further alloying with Cr to form Fe-Ni-Al-Cr has a similar positive effect.
3. In the case of Fe-Cr-Al-Si steels, much smaller total fractions of Al, Si and Cr are required to guarantee adherent diamond film due to a combined effect of these three alloying elements.

These results indicate the feasibility that through composition optimization, especially by adjusting the relative concentrations of the important alloying elements Al, Cr, Si, high quality diamond coatings can be directly fabricated on the designed alloy steels. This will help guide steels design and selection with respect to diamond coatings.

3.1.3. Diamond Nanocone Synthesis on Steels

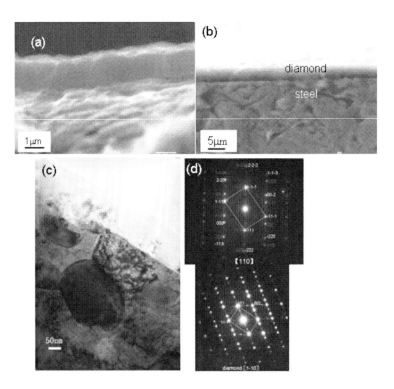

Figure 5. SEM (a,.b) and TEM with SAED (c,d) images of diamond films deposited on two kinds of Al-modified steels.

Without intermediate layer and surface pre-treatment, adherent diamond films with high nucleation density have been directly deposited onto a series of Al-modified steels at a wide temperature range from 370 °C to 740 °C, as shown in Figure 5. More interestingly, well aligned conical diamond nanostructures were also able to be directly fabricated on commercial steels by negatively biasing the substrate and inducing a glow discharge [14]. These diamond nanostructures have demonstrated excellent field electron emission properties in terms of low-field electron emission voltage and high emission current.

3.2. Applying Intermediate Layers

According to the barrier features in obtaining well adhering diamond films on steel substrates, another basic idea to overcome the interfacial problems is to impose additional buffer layers on steel surface to avoid a direct contact between diamond films and steel substrate. This interlayer plays several functions including creating a diffusion barrier between carbon and metallic substrates to prevent carbon diffusion to the bulk, and helping reduce the catalytic effect of base elements by preventing their diffusion to the surface. In addition, the interlayer is expected to provide rich nucleation sites for diamond crystallite growth and enhance the film quality grown on non-diamond substrates. This interlayer can also be designed such that it will significantly improve the adhesion ability of diamond films fabricated on the steel substrates. Depending on its type, physical and chemical nature, an interlayer may remarkably affect the morphology, orientation and purity of the subsequently deposited diamond film.

The use of interlayers to enhance diamond adhesion on steel substrates has been rapidly developed so far, and a variety of pure metals such as Si, Mo, Cr, Ti and their nitrides, oxides and carbides are explored and found desirable interlayer materials [15-18]. It is worthwhile noting that using an aluminum thin film as a high performance interlayer material for diamond growth on steels, hard metals and Ni-base alloys is feasible. The advantages of using Al as interlayer include at least the low materials cost and mature coating technology such as by cladding, PVD sputtering, pack cementation aluminizing.

3.2.1. Diamond Growth on Steel with an Al Interlayer

The feasibility of using Al films as interlayers for growing high quality diamond on stainless steel substrates has been examined using a microwave plasma enhanced CVD reactor. The as-coated Al intermediate film significantly improved the nucleation and phase purity of diamond at early stages of growth on the steel. Graphite phase, which usually grows preferentially on bare steel substrate, was effectively prevented due to this Al interlayer [19]. A direct scratching pre-treatment by diamond suspension was not applicable to this Al-coated steel, as soft Al surface film could be partially removed, resulting in discontinuous diamond films. To overcome this disadvantage, a vacuum interdiffusion annealing process of the Al-coated steel, followed by diamond scratching treatment, enabled the deposition of continuous, dense, and adherent diamond films on Al-interlayered stainless steel substrates [20], as shown in Figure 6.

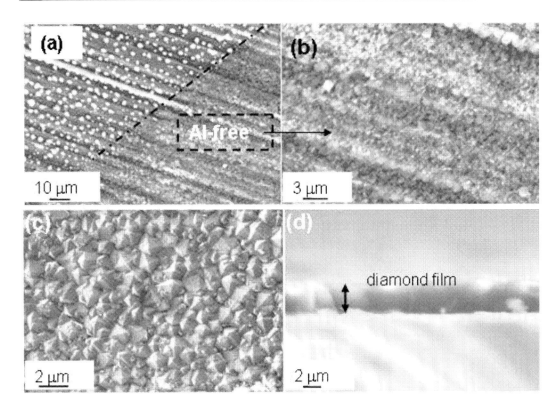

Figure 6. SEM images of diamond films grown for 2 h on steel with patterned Al thin film (boundary shown by dashed line). (a) a general view; (b) a higher magnification view of diamond grown on Al-free region on the steel substrate. (c,d) 14 h growth on Al-covered steel surfaces.

3.2.2. Diamond Growth on Ni-Base Alloys with an Al Interlayer

Normally, for the as-polished blank Ni-base alloy, a surface layer of graphite and amorphous carbon soot form at the early stage of deposition. After prolonged deposition time, the outer-most layer of the deposits consists primarily of diamond particles, which spontaneously peeled off the alloy substrate when exposed to air. Underneath the top layer, a porous intermediate layer of sp^2-bonded carbon (graphite) exists. These results demonstrate that graphite layer was preferentially formed on the alloy substrate, then diamond gradually nucleates and grows on this intermediate layer. The spontaneous de-bonding of the diamond film from the alloy substrate is primarily caused by the poor adhesion strength at the diamond film-graphite layer interface. The preferential formation of this intermediate layer (graphite/amorphous carbon) over diamond is again closely associated with the strong catalytic activity of the base metal Ni. Obviously, in order to enhance the adhesion ability of diamond on this type of transition metal substrate, formation of the detrimental intermediate layer must be avoided.

Figure 7 shows the diamond grown on the alloy that has been pre-coated with an Al interlayer. The diamond crystallites are well faceted, densely packed and form a continuous film that sticks strongly to the alloy substrate. When the top diamond film is mechanically removed by cyclically bending the sample, the exposed alloy surface exhibits metallic luster. The diamond characteristic peak in Raman spectrum is broadened and shows a significant upward shift (from standard 1332 to 1342 cm^{-1}) due to the accumulated compressive stress in the film, while the Raman spectrum of the as-exposed alloy substrate shows no evidence of carbon-related information, indicating a clear diamond/alloy interface structure after using Al interlayer material. Adhesion ability of the diamond film was evaluated by Rockwell C indentation test and the results are shown in Figure 8. Even under a high load up to 1470 N, no significant film spallation occurs around the imprint except for some local cracks. Figure 9 compares the Raman spectra changes on different surface regions of the indented diamond film. In the regions far away from the indentation crater and the cracks, where diamond films still remain continuous and well adherent, the diamond characteristic peak is still centered approximately at 1343 cm^{-1} (I), an indication of the high residual compressive stress in the film associated with its strong bonding to the alloy substrate. However, the diamond characteristic peak measured around the cracked regions is remarkably sharpened and approaches standard diamond peak position 1332 cm^{-1} (II). This can be attributed to a significant stress release in the diamond film after its debonding to the substrate. This comparison indicates clearly that the Al interlayer has inhibited the formation of detrimental graphite at the diamond-substrate interface [21].

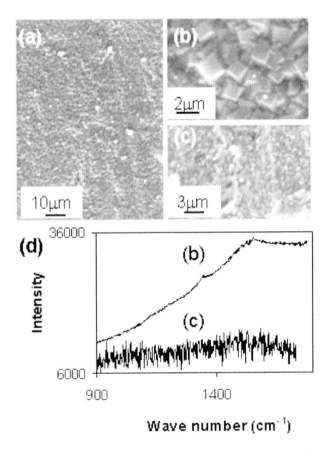

Figure 7. SEM images of diamond film deposited for 6 h on Ni-base Inconel alloy pre-coated with an Al interlayer. (a): a general view; (b): a magnified view of the top film surface; (c) as-exposed alloy substrate after removing top diamond film; (d) the corresponding Raman spectra of diamond film and the exposed substrate surface.

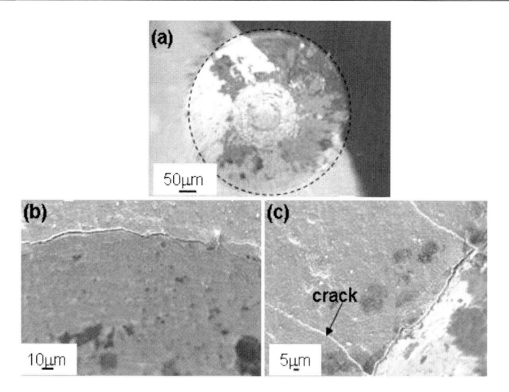

Figure 8. SEM images of diamond films after Rockwell C indentation test with a load of 1470 N. (a) a general view; (b,c) magnified views around the imprint.

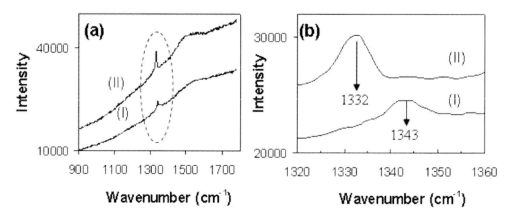

Figure 9. Raman spectra of diamond films after Rockwell C indentation test with a load of 1470 N from figure 3. The magnified spectra of the circled region of (a) are presented in (b). (I) measured from adherent regions far away from cracks and imprint; (II) measured from around the cracks.

3.2.3. Diamond Growth on WC-Co with an Al Interlayer

Diamond deposition on WC-Co cutting tool inserts with the aids of several intermediate layers has been conducted. The results show that the diamond films directly grown on the as-received WC-Co insets are of poor quality in terms of purity, nucleation density, and adhesion strength, as shown in Figure 10a-b. The diamond films grown on the inserts pre-coated with TiN, TiCN intermediate layers demonstrate higher phase purity but slow nucleation rates.

Severe wrinkles and spontaneous peeling off the substrates occur in the diamond films after cooling to room temperature. When an aluminum thin film was deposited either directly adjacent to WC-Co substrate, or on top of TiN, TiCN intermediate layers, the nucleation density and adhesion ability of diamond films on the tool inserts are significantly improved (Figure 10c) [22].

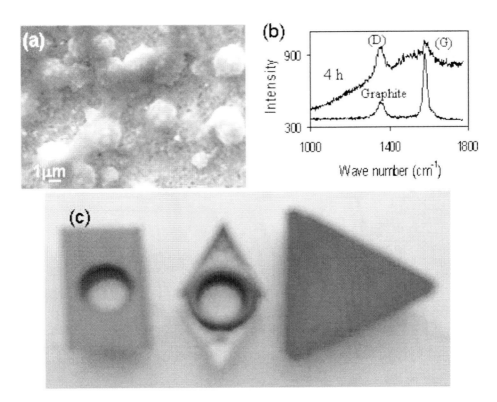

Figure 10. SEM image (a) and Raman spectrum (b) of diamond film deposited for 4 h on blank WC-Co substrate. (c) continuous diamond coatings deposited with an Al interlayer.

3.2.4. Diamond Growth on Steel with an Ultra-Thin W-Al Interlayer

The adherence of diamond coated on steel is usually required to be strengthened by applying thick intermediate layers. Ideal interlayer materials should offer high diamond nucleation density, matching thermal expansion coefficient and enhanced interfacial adhesion. However, diamond nucleation density/rates on the intermediate layer materials currently in use are not sufficiently high, and the involving multi-interfacial adhesion issues (substrate/interlayer, interfacial reaction layer, interlayer/diamond film) for thick interlayer (several μm and higher), need to be strengthened. An ultrathin single interlayer is also associated with problems like insufficient barrier effect for obtaining high purity diamond. Our most recent findings have indicated that elemental Al has a unique effect on inhibiting graphitization of carbon on transition metals, while the strong carbide-forming element, W, can greatly enhance diamond nucleation leading to improved surface smoothness. We thus developed a nano-scale W-Al dual metal interlayer to facilitate deposition of continuous, adherent and smooth diamond thin films on steels.

Figure 11 shows the SEM and AFM images, Raman spectrum, synchrotron C k-edge NEXAFS spectrum recorded in total electron yield (TEY), and GIXRD patterns of the nanocrystalline diamond film grown on W-Al interlayered steel, respectively. Only one diamond characteristic peak centered at 1342 cm^{-1} appears in the spectrum along with high background intensity caused by fluorescence. The upward shift of the peak position from standard diamond at 1332 cm^{-1}, is attributed to a compressive stress accumulation in the film, and it is also a sign of good adhesion of the film to the substrate. In fact, once the diamond film delaminates from the substrate, the diamond peak measured from the free-standing films is located at 1332 cm^{-1} again due to stress release after de-bonding.

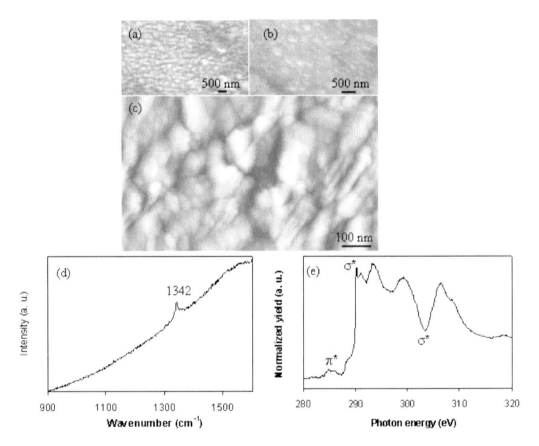

Figure 11. SEM (a, b) and AFM (c) images of diamond thin film grown for 4 h (a) and 7 h (b, c) on 50 nm W-Al interlayered steel substrate. Raman spectrum (d), TEY NEXAFS spectrum (e) of diamond film grown for 7 h on 50 nm W-Al interlayered steel substrate.

The significant broadening of the peak can be caused by both the compressive stress and the refinement of the diamond grain size in a nano-scale, while the absence of any obvious graphitic peaks indicates the high purity of the diamond film synthesized. This is further confirmed by carbon K-edge synchrotron NEXAFS spectrum. The spectrum demonstrates a sharp absorption edge at 289.7 eV and a huge dip at 303.2 eV, corresponding to featured σ bonding of pure diamond. The non-diamond carbon peak at 285.8 eV associated with sp^2 structured π bonding, is very weak in the spectrum.

The bonding ability of the diamond film to the steel is evaluated by Rockwell C indentation test with a high load of 1470 N (Figure 12). No obvious delamination or cracking of the diamond films is observed around the imprint. The diamond film-coated steel plate is also sheared with a metal cutting machine. An approximately 21° bending deformation of substrate occurs along the cutting edges, and the diamond film coated is accordingly subjected to a huge shear stress. Even so, no large area film spallation is induced. Although lateral cracks appear within the film in the shear zone, the diamond film still sticks well to the substrate at the cutting edge. These results confirm that the diamond film coated on steel possesses high adhesion strength.

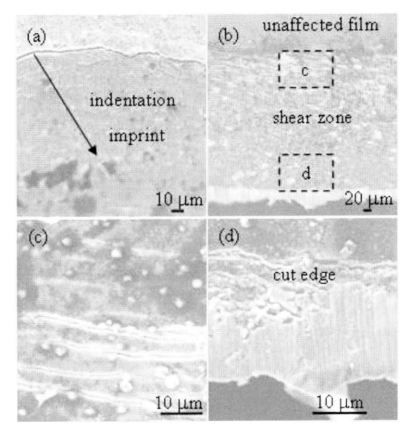

Figure 12. SEM surface images of diamond grown for 7 h on 50 nm W-Al interlayered steel. (a): after 1470 N Rockwell C indentation test; (b): general view after cutting of diamond-coated steel sheet, (c): magnified view of shear deformation zone and (d): magnified view of cutting edge. No significant film spallation occurs after applying high mechanical loads.

In summary, a W-Al ultrathin interlayer combination with a total thickness of 50 nm has produced promising synergic effect on enhancing diamond nucleation, adhesion and surface smoothness grown on stainless steel substrates. These improvements can be attributed to the inward diffusion of Al to form Fe-Al compounds inhibiting graphitization and substrate metal diffusion, while surface carburization of W facilitating diamond nucleation and providing stronger chemically bonded interfaces [23].

4. MECHANISM ON IMPROVED DIAMOND QUALITY ON AL-MODIFIED TRANSITION METALS

The transition metals Fe, Co, Ni have been known to be active with the carbonaceous precursors and graphitic phase is catalytically activated. Fe also significantly consumes carbon because of a large diffusion of carbon in it. Therefore, the diamond nucleation density on conventional Fe-base alloys is low and a long incubation period for nucleation is required. In contrast, element Al has been proved to remarkably decrease the catalytic deactivity of the transient metals Ni and Fe, and to inhibit the trend for graphite formation. Furthermore, Al imposes a barrier effect for the inward penetration of carbon due to decreased carbon solubility in the presence of Al. As a direct consequence, a sufficient carbon source supply is available for diamond nucleation, even at an early stage of CVD process. The high initial nucleation density helps increase the effective contact points and bonding strength between the film and the substrate interface. Our recent study reveals that adherent nanocrystalline diamond films can be directly deposited on a kind of Fe-9Cr steel alloyed with a small fraction of Al. The variation of the diamond growth mode from micro- to nano-crystallite dependent on the steel compositions is believed to be caused by a delicate balance between the roles played by Fe and Al [13].

5. DIAMOND GROWTH ON TI AND ITS ALLOYS

Metallic Ti and its alloys are promising materials widely used in aerospace, bio-medical and chemical industries because of their low density, high specific strength and high corrosion resistance. However, their extended applications are limited by the poor tribological properties [24,25,26]. Chemical vapor deposition of wear/corrosion resistant diamond coatings on such substrates will significantly enhance the durability and service performances of these materials. Adherent diamond coatings are difficult to deposit on the Ti metal and alloys due to the large mismatch of their thermal expansion coefficients. In addition, severe chemical reaction between the gas reactants and the base materials strongly alter the microstructures of the substrate materials. For instance, atomic hydrogen is usually considered crucial to synthesize high quality diamond during the chemical vapor deposition process, as it preferentially etches non-diamond carbon species. However, hydrogen itself easily dissolves into Ti substrate and induces phase transformation and microstructure changes, which may deteriorate the mechanical properties of the substrates [27-30]. Although the formation of an intermediate titanium carbide layer is usually considered beneficial to enhance diamond nucleation and adhesion, the introduction of such a new hetero-interface also complicates the adhesion failure models of diamond coatings.

Up to now, extensive investigations have been conducted to exploit new methods to produce well adherent diamond coatings on Ti metal and alloys of different shape and dimensions. However, obstacle still exists in obtaining smooth, adherent diamond coatings without altering the microstructure and properties of the base materials. Almost without exception, the hydrogen ratios used in those previous studies are very high. In this paper, the influence of an extremely high CH_4 concentration up to 100% on the nucleation, growth and adhesion behaviors of diamond coatings on pure titanium substrate will be introduced. To decrease the thermal stress effect, a moderate deposition temperature is adopted. The structure changes of the base material after plasma exposure are also characterized by synchrotron fluorescence spectroscopy and micro-diffraction.

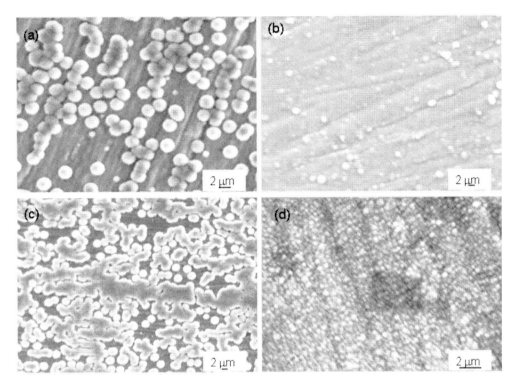

Figure 13. SEM images of diamond nucleation and growth on Ti. (a) as polished substrate under CH4-99%H2 for 10 hrs, (b) diamond suspension pre-scratched substrate under CH4-99%H2 for 10 hrs, (c) as polished substrate under 100%CH4 for 2 hrs, (d) diamond suspension pre-scratched substrate under 100%CH4 for 10 hrs.

Figure 13 shows the nucleation and growth morphologies of diamond on Ti substrate with different surface conditions and deposition parameters. For the as-polished Ti substrate which means no pre-scratching with diamond suspension is applied, the nucleation density and nucleation rate with a 1%CH$_4$ concentration are very low. After 10 hrs of deposition, the diamond crystals are only sparsely distributed on the Ti substrate surface. The diamond crystallites are not as well faceted as those grown on Si wafer under the same deposition conditions. This can be attributed to a significant diffusion of carbon into the Ti substrate which causes a lack of carbon oversaturation on the top surface of Ti, making rapid diamond nucleation and growth more difficult. When a diamond suspension scratching pretreatment is applied, the diamond nucleation and growth rates are markedly increased. After 10 hrs deposition, the diamond crystals have coalesced quickly to form a dense and continuous diamond coating. The reason for the significant increase in nucleation and growth rates of diamond is normally ascribed to a pre-seeding effect. Without such a scratching pre-treatment, when a 100%CH$_4$ is used, the nucleation density and growth rates of diamond on Ti substrate still increase dramatically. Nonetheless, a disadvantage of using an extremely high CH$_4$ ratio is that the diamond quality may be deteriorated, due to a weak etching effect of non-diamond carbon species by hydrogen plasma. The three step deposition process with alternative CH$_4$ concentrations also produced dense, continuous and smooth diamond coatings.

Figure 14 shows the X-ray absorption spectra of Ti L- and K-edge measured from Ti surfaces subjected to different treatments reveal that the hydrogen plasma etching has significantly changed the chemical states on Ti surfaces in terms of the shifts in peaks position and intensities. Its effect on the substrate microstructures will be introduced in the following work.

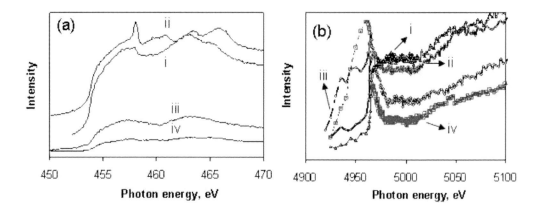

Figure 14. Ti L-edge (a) and K-edge (b) XANES spectra measured from surfaces of: (i) as-polished Ti, (ii) air oxidized Ti, (iii) Ti etched in single H2 plasma, (iv) exposed Ti substrate after external diamond coating delamination.

To evaluate the adhesion of diamond coatings grown on the Ti substrate, Rockwell C indentation tests are conducted under a load of 150 kg and the results as shown in Fig. 15. Although the diamond coatings synthesized with 1%CH$_4$ show a dense and continuous appearance, after indentation, a severe spallation of diamond coatings around the indenter imprint occurs, indicating the interfacial bonding strength is actually very weak in nature. Comparatively, no significant spallation occurs to the diamond coatings grown with a 100%CH$_4$ or alternative CH$_4$ concentrations. Instead, many longitudinal cracks, initiated by the high load, are initiated on the coating surface due to the brittle nature of the diamond coatings deposited. These results indicate that the diamond coating deposited at a higher CH$_4$ concentration demonstrates higher adhesion strength. The benefit may come from two aspects. One factor is that the rapid nucleation density and rate of diamond can effectively increase the contact area of diamond coating with the underlying substrate by reducing interfacial porosity. The other beneficial factor is that an abundant carbon source supply, like in the case of deposition under high CH$_4$ concentrations, may promote a simultaneous formation of titanium carbide. Once a quick coverage of the substrate surface by diamond and TiC is facilitated, it acts as a diffusion barrier to prevent continued carbon and hydrogen penetration. This assumption is confirmed by a close SEM observation. After mechanical bending deformation of the diamond-coated Ti substrate, some fresh substrate surfaces are exposed due to local diamond delamination. For diamond coatings deposited under 1%CH$_4$, the adhesion failure is primarily observed at the titanium carbide-substrate interface. For diamond coatings deposited under 100%CH$_4$, the coating debonding occurs both at the diamond-carbide layer interface and at the carbide layer-substrate interface. The results indicate that the properties of the intermediate carbide layer, such as its thickness and porosity depending on the deposition conditions, significantly influence the adhesion ability of diamond coatings grown on it. As the thermal expansion coefficient of TiC lies between those of diamond and the substrate, theoretically, a thicker carbide intermediate layer produced under higher CH_4 concentrations may help reduce the thermal stress and enhance the adhesion of diamond coating, as demonstrated in the indentation test. However, the coating debonding occurring at different interfaces indicates that further optimization of deposition conditions thereby a proper control of the interface properties is still required in order to develop well-adherent diamond coatings on such strong carbide-forming substrates like titanium.

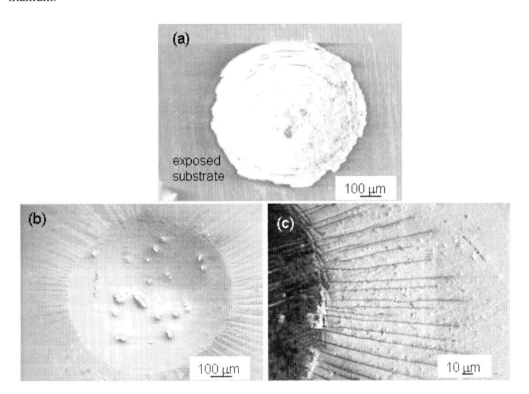

Figure 15. SEM surface images of diamond-coated Ti substrates after Rockwell indentation test. Diamond coatings synthesized using (a) CH4-99%H2, (b-c) 100%CH4.

It is well known that there is usually a significant hydrogen interstitial diffusion problem during diamond deposition on Ti and its alloys. Accordingly, the influence of hydrogen plasma on the structure changes of underlying Ti substrate is further studied. After an exclusive etching of the Ti substrate by hydrogen plasma, synchrotron-based X-ray near edge absorption spectroscopy, fluorescence spectroscopy and Laue micro-beam diffraction measurements are conducted, and some preliminary results are briefly introduced in the following. First, a cross section of the hydrogen plasma etched Ti specimen is prepared for a two-dimensional synchrotron X-ray analysis, while the fluorescence element mapping of Ti is basically used to determine the exact location of the specimen edge for a subsequent depth resolution. Fig. 16 shows the quite different Laue diffraction images measured from the specimen surface to four hundred micrometers deep inside the substrate, respectively. Most of the Laue spots from the images can be correctly indexed by hcp α–Ti, but titanium hydride TiH$_2$ patterns can not be detected across the entire depth measured. However, the lattice orientation map derived from the Laue diffraction images clearly reveals that there is a significant microstructure coarsening on the near surface region of Ti substrate after hydrogen plasma etching. The pole figures orientation mapping of the measurement region indicates an obvious indication of preferential orientation along [100]. As the hydrogen-induced microstructure coarsening is normally accompanied by significant deterioration of the mechanical properties of the base material, like decreased fatigue life and impact strength, it should be mitigated as low as possible. In current study, no significant microstructure coarsening is observed on Ti substrate after exposure to 100% CH$_4$ plasma (results not shown here), and this can be attributed to a barrier effect provided by the rapidly formed diamond and TiC surface coverage. These results indicate that high CH$_4$ concentrations used for diamond coating on Ti materials have the advantages of both enhancing coating adhesion and mitigating substrate damage.

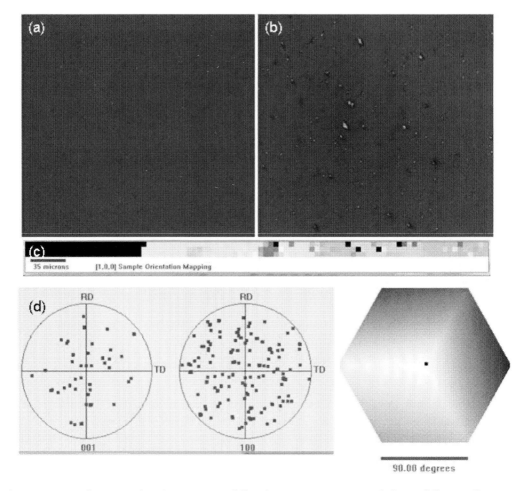

Figure 16. Synchrotron micro-beam Laue diffraction patterns measured from different distances beneath the Ti substrate surface. (a) outer region close to surface, (b) inside the substrate. (c) [100] orientation map derived from Laue diffraction.

CONCLUSION

The nucleation, growth, and adhesion properties of diamond synthesized on transition metal substrates by chemical vapor deposition method are strongly dependent on both the substrate compositions and the deposition conditions. Diamond synthesis on the conventional carbon steel, stainless steel, Ni-base alloys, and hard metals are usually difficult due to the low nucleation rate, low nucleation density and poor bonding strength to the substrates. By modifying the chemical and physical properties of the bulk or the near surface area of substrate materials, such as by means of alloying design and surface modification, rich sites for diamond nucleation can be created and the film adhesion strength with the substrate is enhanced. The following conclusions can be summarized:

1. Strong carbide-forming alloying elements Si, Cr, Ti, Mo and Mn, individually added into steels insignificantly improve the quality of diamond films subsequently produced. The diamond films contain high portions of non-diamond phases and they spontaneously peel off the substrates after CVD process.
2. Continuous and adherent diamond films can be successfully fabricated on Al-modified model alloy steels or commercial alloy steels, even without applying intermediate layer any nucleation enhancing surface pre-treatment. Various diamond nano-structures have been directly fabricated on such Al-modified steel substrates of complex shapes, in a wide temperature range.
3. Due to its great cost advantage and mature coating technique and efficiency, Al is confirmed to be a promising intermediate layer material for diamond fabrication on the ferrous metal substrates.
4. The beneficial role Al plays on the diamond growth on transition metals Fe, Co, Ni is primarily associated with its delicate competition and balance with the base metal (Fe, Co, Ni) in terms of their activation and deactivation ability on catalyzing graphite formation during deposition process.
5. A high CH_4 concentration used for diamond coating on Ti and its alloys makes it possible to deposit continuous and adherent nanocrystalline diamond coatings at moderate temperatures. Accordingly, substrate damage problem such as grain coarsening is significantly mitigated.
6. Synchrotron radiation technology shows a powerful tool to characterize the diamond quality, diamond-substrate interfacial chemistry and the structural changes of the underlying metal substrates.

ACKNOWLEDGMENT

This study is sponsored by the Canada Research Chair Program and by the Natural Sciences and Engineering Research Council of Canada (NSERC). The synchrotron data were measured from the 11ID-1, 10ID-2, 06B1-1 and 07B2-1 beamlines, the Canadian Light Source, which is supported by the NSERC, NRC and the University of Saskatchewan.

REFERENCES

[1] May, P. W. Phil. T. Roy. *Soc. A.* 2000, 358, 473-495.
[2] Xu, N. S.; Huq, S. E. *Mater Sci engineer R.* 2005, 48, 47-189.
[3] Jiang, X., Fryda, M., Jia, C. L. *Diamond Relat. Mater.* 2000, 9, 1640-1645.
[4] Yan, B. B. *Surf. Coat. Technol.* 1999, 115, 256-265.
[5] Ali, N.; Fan, Q. H.; Gracio, J.; Ahmed, W. *Surf. Engineer.* 2002, 18, 260-264.
[6] R.J. Narayan, *J. Adhes. Sci. Technol.* 2004, 18, 1339-1365.
[7] Maguire, P. D.; McLaughlin, J. A. *Diamond Relat. Mater.* 2005, 14, 1277-1288.
[8] Chen, X.; Narayan, J. *J. Appl. Phys.* 1993, 7, 4168-4173.
[9] Spinnewyn, J.; Nesladek, M.; Asinari, C. *Diamond Relat. Mater.* 1993, 2, 361-364.
[10] Davanloo, F.; Park H.; Collins, C. B. *J. Mater. Res.* 1996, 11, 2042-2050.
[11] Li, Y. S.; Yang, Q.; Xiao, C.; Hirose, A. *Thin Solid Films* 2008, 516, 3089-3093.
[12] Li, Y. S.; Hirose, A. *Surf. Coat. Technol.* 2007, 202, 280-287.
[13] Li, Y. S.; Hirose, A. *Chem. Phys. Lett.* 2006, 433, 150-153.
[14] Li, Y. S.; Pan, T.J.; Tang, Y.; Yang, Q.; Hirose, A. *Diamond Relat. Mater.* 2011, 20, 187-190.
[15] Buijnsters, J.G.; Shankar, P.; Gopalakrishnan, P. *Thin Solid Films.* 2003, 426, 85-93.
[16] Haubner, R.; Lux, B. *Int. J. Refract. Met. Hard. Mater.* 2006, 24, 380-386.
[17] Lin, C. R.; Kuo, C.T. *Diamond Relat. Mater.* 1998, 7, 903-907.
[18] Li, Y. S.; Xiao, C.; Hirose, A. *J. Am. Ceram. Soc*, 2007, 90, 1427-1433.
[19] Li, Y. S.; Tang, Y.; Yang, Q.; Hirose, A. *Materials Chemistry and Physics*, 2009, 116, 649-653.
[20] Li, Y. S.; Tang, Y.; Yang, Q.; Hirose, A. *International Journal of Refractory Metals and Hard Materials*, 2009, 27, 417-420.
[21] Li, Y. S.; Hirose, A. *Applied Surface Science*, 2008, 255, 2251-2255.
[22] Li, Y. S.; Tang, Y.; Yang, Q.; Shimada, S.; Wei, R.; Lee, K.Y.; Hirose, A. *International Journal of Refractory Metals and Hard Materials*, 2008, 26, 465-471.
[23] Li, Y. S.; Tang, Y.; Yang, Q.; Maley, J.; Sammynaiken, R.; Regier, T.; Xiao, C.; Hirose, A. *ACS Appl. Mater. Interfaces*, 2010, 2, 335–338.
[24] N.A. Braga, C.A.A. Cair, J.T. Matsushima, M.R. Baldan and N.G. Ferreira, *J. Solid State Electrochem.*, 14 (2010) 313.
[25] T. Grogler, E. Zeiler, A. Franz, O. Plewa, S.M. Rosiwal and R.F. Singer, *Surf. Coat. Technol.*, 112 (1999) 129.
[26] K.G. Budinski, *Wear.*, 151 (1991) 203.
[27] T. Grogler, E. Zeiler, M. Dannenfeldt, S.M. Rosiwal and R.F. Singer, *Diamond Relat. Mater.*, 6 (1997) 1658.
[28] T. Grogler, E. Zeiler, A. Horner, S.M. Rosiwal and R.F. *Singer, Surf. Coat. Technol.*, 98 (1998) 1079.
[29] X.L. Peng and T.W. Clyne, *Thin Solid Films*, 293 (1997) 261.
[30] Y.Q. Fu, B. Yan and N.L. Loh, *Surf. Coat. Technol.*, 130 (2000) 173.

INDEX

A

absorption spectroscopy, 213
acetone, 116
acid, 6, 8, 10, 13, 18, 19, 24, 30, 76, 79, 81, 82, 87, 159, 160, 177, 181
activated carbon, 194
activation energy, 38
active site, 82
adamantane, 187
adenine, 3, 9, 13
adhesion, x, 30, 38, 68, 116, 119, 163, 165, 178, 188, 195, 196, 198, 199, 200, 202, 203, 204, 206, 207, 208, 209, 210, 212, 214, 215
adhesion properties, 199, 215
adhesion strength, 197, 200, 204, 206, 209, 212, 215
adsorption, 2, 3, 4, 10, 12, 20, 21, 82, 158, 159, 161, 166, 170, 178, 179, 181, 186
aerospace, 210
AFM, 75, 77, 187, 208
AIDS, 5
albumin, 168
alcohols, 86, 88
aliphatic compounds, 78, 88
alkenes, 159
allele, 25
allergy, 176
allylamine, 16
amine, vii, 1, 3, 16, 17, 22, 159, 177, 178
amine group, 16, 177
amines, 4, 160
amino, 4, 5, 8, 16, 17, 158, 160, 161, 187
amino acid(s), 4, 5, 8, 160, 187
amino groups, 16, 161
ammonia, 159, 160, 174
ammonium, 11
amplitude, 40, 108
anchorage, 176

angioplasty, 192
aniline, 80
annealing, 38, 120, 123, 124, 127, 145, 148, 149, 151, 168, 175, 203
antibiotic, 169
antibody, 166
antigen, 166
antioxidant, 12, 171
apoptosis, 169, 171
aqueous solutions, 75, 78
argon, 47, 56, 163, 173
aromatic compounds, 78, 88
aromatics, 78, 80, 83, 88
ascites, 182
ascorbic acid, 7, 24, 25, 26, 27
assessment, 188
atmosphere, 3, 10, 56, 57
atmospheric pressure, 174, 190, 196
atomic force, 75
atomic orbitals, vii, 29, 31, 32
atoms, ix, 32, 33, 34, 35, 38, 51, 116, 118, 132, 133, 142, 145, 146, 147, 148, 149, 150, 151, 161, 172, 173
attachment, x, 155, 164, 165, 171, 178, 179, 182
auditory cortex, 167
axons, 164

B

background noise, 167
bacteria, 166, 182
band gap, 36, 37, 38, 43, 44, 47, 48, 49, 51, 52, 132, 175
barriers, x, 191, 195
base, 22, 45, 47, 101, 159, 198, 199, 202, 203, 204, 205, 210, 214, 215
beams, 44, 147, 153
behavioral change, 21

behaviors, 186, 210
bending, 204, 209, 212
beneficial effect, 200
benzene, 33, 36, 80, 83, 84, 85
bias, 163, 186, 188
binding energy, 32, 172
biochemical processes, 9
biocompatibility, vii, x, 1, 2, 18, 155, 163, 167, 168, 171, 175, 176, 177, 183, 184, 185, 188, 193
biocompatible materials, 185
biological activity, 9, 161, 166, 170, 177
biological fluids, 20
biological systems, 5, 22, 168
biomarkers, 168, 190
biomass, 183
biomaterials, x, 155, 163, 188
biomedical applications, viii, 29, 37, 156, 158, 169
biomolecules, vii, x, 1, 2, 3, 10, 15, 16, 20, 22, 155, 156, 161, 177
biosensors, vii, ix, 1, 2, 3, 4, 7, 14, 16, 17, 18, 22, 26, 155, 156, 157, 158, 159, 165, 166, 177, 180, 184
biotechnology, 165, 167
biotin, 161, 168
bisphenol, 16, 86
bloodstream, 182
bonding, vii, 16, 18, 29, 31, 32, 33, 34, 36, 38, 49, 61, 116, 121, 146, 204, 205, 208, 209, 210, 215
bonds, 31, 32, 33, 36, 37, 49, 51, 133, 146, 156, 166
bone, 169, 179, 184, 188
bone marrow, 179, 188
boric acid, 132
bottom-up, 180
brain, 27, 167
Brazil, 73, 90, 131
breakdown, 176
breast cancer, 181
brittle nature, 212
building blocks, vii, 29
bulk materials, 197
by-products, 83, 166

C

C reactive protein, 166
cadmium, 157, 181
calcium, 193
calibration, 11, 17
cancer, 169, 181, 189
candidates, 22, 151
capillary, 9, 20, 26
carbides, 197, 198, 203
carbon atoms, 30, 35, 36, 38, 61, 94, 156, 161, 172, 173, 181

carbon dioxide, 77, 161, 175
carbon emissions, 175
carbon film, 34, 39, 50, 157, 172, 173, 174, 190, 191, 192
carbon materials, 2, 24, 32, 61, 161, 173, 186, 193
carbon nanotubes, 2, 23, 26, 146, 168, 180, 186
carbon tetrachloride, 159
carboxyl, vii, 1, 3, 7, 15, 18, 19, 22, 24, 160
carboxylic acid, 86, 88
carboxylic groups, 15, 16, 18
carcinoma, 182
catalysis, 14
catalyst, 15, 62, 63, 198
catalytic activity, 2, 14, 120, 157, 171, 204
catalytic effect, 21, 203
catecholamines, 20
cation, 88
C-C, 6, 133, 159
cell biology, 188
cell culture, 163, 179
cell death, 169
cell division, 189
cell line, 170, 178
cell lines, 178
cell membranes, 170
central nervous system (CNS), 164
ceramic(s), ix, 94, 120, 121, 127, 131
charge coupled device, 116
chemical, vii, viii, ix, x, 2, 3, 10, 16, 17, 20, 21, 29, 30, 31, 32, 34, 36, 37, 45, 52, 55, 56, 57, 58, 60, 61, 62, 63, 64, 66, 68, 70, 74, 75, 76, 78, 79, 88, 90, 94, 95, 114, 125, 131, 132, 135, 146, 155, 156, 159, 160, 162, 163, 165, 167, 172, 173, 174, 175, 177, 180, 181, 183, 184, 187, 192, 194, 196, 203, 210, 212, 215
chemical bonds, 31, 32, 135
chemical characteristics, 2
chemical inertness, viii, 20, 29, 30, 37, 74, 125, 175
chemical kinetics, 61
chemical properties, vii, x, 29, 31, 52, 94, 114, 155, 163, 177, 181, 192
chemical reactions, 58, 62, 64
chemical reactivity, 68
chemical stability, x, 90, 156, 167, 183
chemical structures, 75
chemical vapor deposition, ix, 2, 3, 45, 74, 131, 132, 187, 196, 210, 215
chemical vapour deposition, viii, 21, 34, 55, 172, 173
chlorine, 159, 160
chromatograms, 13
chromium, 119, 198
cladding, 203
classes, 148, 157, 196

Index

cleaning, 46, 47, 61
clinical trials, 176
cluster model, 36, 47, 51
clustering, 35, 36
clusters, vii, ix, 15, 29, 30, 35, 36, 39, 48, 51, 52, 122, 131, 135, 138, 142
CO2, 62, 63, 75, 76, 83, 89
coatings, x, 38, 39, 96, 106, 114, 116, 118, 119, 120, 157, 165, 166, 172, 173, 174, 175, 177, 184, 185, 190, 192, 193, 195, 196, 200, 202, 207, 210, 212, 213, 215
cobalt, 198
collagen, 165
collisions, 148, 150
color, 97, 134
combined effect, 202
combustion, 121
commercial, 23, 74, 77, 121, 122, 123, 124, 127, 203, 215
compatibility, 192, 196
competition, 215
compilation, vii
complement, 74
complex numbers, 44
composites, viii, 55, 57, 59, 63, 68, 69, 132, 184
composition, viii, 34, 60, 65, 93, 111, 114, 121, 122, 165, 172, 197, 202
compounds, 7, 9, 10, 14, 21, 75, 80, 88, 146, 159, 180, 196, 209
compressibility, 196
computation, 151
computer, 76
conduction, 36, 38, 43, 44, 60
conductivity, viii, 29, 30, 35, 37, 38, 52, 72, 75, 111, 120, 135, 164, 196
configuration, 31, 36, 47, 146, 147
conjugation, x, 156, 167
construction, vii, 1, 2, 16, 22, 26, 165
consumption, 9, 65
contour, 66
control group, 134
cooling, 60, 168, 198, 207
coordination, 116
copolymer, 25
copper, 7, 96, 161, 198
corrosion, x, 119, 155, 177, 196, 197, 210
cortex, 167
cortical bone, 169
cortical neurons, 179
cost, x, 37, 56, 90, 103, 114, 156, 174, 195, 197, 203, 215
covalent bond, 17, 32, 166
covalent bonding, 17

cracks, 66, 94, 205, 206, 209, 212
critical value, 64
crystal growth, 162
crystal structure, 161
crystalline, viii, ix, 32, 33, 34, 55, 57, 58, 63, 69, 116, 120, 131, 145, 146, 148, 149, 150, 151, 196
crystallinity, ix, 145, 148
crystallites, 204, 211
crystals, 106, 116, 117, 118, 122, 124, 146, 149, 211
cues, 179
culture, 165, 171
culture medium, 171
CV, 77, 81, 82
CVD, v, vi, viii, ix, 45, 55, 56, 57, 63, 68, 69, 70, 71, 72, 74, 75, 93, 94, 96, 97, 100, 104, 106, 107, 109, 110, 111, 112, 114, 118, 125, 131, 132, 133, 134, 135, 138, 141, 142, 162, 163, 189, 195, 196, 198, 199, 203, 210, 215
cyanide, 77
cycles, 80, 83, 86, 166, 178
cycling, 7, 15
cysteine, 10, 25
cytochrome, 17, 18, 23, 24, 27, 186
cytochromes, 9
cytocompatibility, 193
cytokines, 165, 177
cytometry, 169, 171
cytoplasm, 169, 181
cytosine, 13
cytotoxicity, x, 156, 157, 165, 170, 184, 185, 190, 192

D

decomposition, 80, 83, 86, 118, 171
defects, vii, ix, 4, 50, 64, 131, 132, 133, 145, 147, 149
deficiency, 5, 57
deformation, 209, 212
degradation, 30, 38, 75, 161, 165, 179
degree of crystallinity, 149
denaturation, 159, 165
deposition, vii, viii, ix, x, 14, 15, 20, 37, 39, 45, 46, 47, 75, 86, 96, 113, 114, 115, 116, 117, 118, 119, 120, 121, 122, 124, 127, 131, 132, 133, 134, 137, 138, 141, 142, 156, 162, 163, 167, 173, 174, 175, 179, 190, 193, 195, 196, 198, 199, 203, 204, 206, 207, 210, 211, 212, 213, 215
deposition rate, 37, 45, 119
deposits, 106, 116, 120, 124, 204
depth, 149, 150, 151, 213
derivatives, 8, 23, 24, 156, 187
desorption, 5, 16, 170

detection, vii, 1, 2, 3, 4, 5, 6, 7, 8, 9, 10, 11, 12, 13, 14, 15, 17, 18, 19, 20, 22, 23, 24, 25, 26, 27, 165, 166, 167, 168, 189
detonation, 159, 160, 161, 167, 186, 187
diabetes, 7, 27
diamonds, vii, ix, 69, 70, 72, 113, 116, 166, 189
dielectric constant, 38, 120
dielectric strength, viii, 29, 30, 37, 175
dielectrics, 38
differential equations, 148
diffraction, ix, 60, 61, 116, 131, 137, 142, 211, 213, 214
diffusion, ix, 12, 18, 45, 56, 60, 62, 70, 83, 113, 118, 119, 120, 132, 196, 198, 203, 209, 210, 211, 212, 213
diffusion process, 70
diffusivity, x, 65, 195
dipoles, 10, 11
directionality, 156
discharges, 56
diseases, 9, 11, 166
disinfection, 74
disorder, 36, 37, 47, 48, 49, 50, 116
dispersion, 41, 161
dissociation, 5, 150
distribution, 15, 67, 101, 133, 164
diversity, 31
DNA, 9, 10, 24, 26, 156, 157, 159, 165, 168, 172, 178, 185, 186, 189
DNA damage, 189
DNA repair, 168
Doha, 73
dopamine, 23, 24, 25, 26
dopants, 38, 147
doping, ix, 18, 30, 38, 50, 51, 131, 132, 133, 135, 140, 141, 147, 148, 151, 156, 157, 178
drawing, 94, 95
drinking water, 74
drug carriers, 157
drug delivery, x, 155, 157, 158, 161, 169
drugs, 169, 172
drying, 174
durability, 118, 210
dyes, 75, 79, 80, 158, 168

E

ECM, 163, 171
electric field, 165
electrical conductivity, 38, 52
electrical properties, vii, 29, 30, 38, 167
electrocatalysis, 15, 112
electrochemical behavior, 5, 15, 18, 23, 75, 79, 83

electrochemical deposition, 14, 174
electrochemistry, 4, 10, 14, 18, 22, 24, 27, 74, 75
electrode surface, 7, 9, 12, 14, 16, 17, 18, 19, 22, 74, 80, 83, 86
electrodeposition, 14, 15
electrodes, vii, 1, 2, 3, 4, 5, 6, 7, 8, 9, 10, 11, 12, 13, 14, 15, 16, 17, 18, 19, 20, 21, 22, 23, 24, 25, 26, 27, 45, 74, 77, 80, 83, 86, 87, 89, 112, 166
electrolysis, 75, 76, 83, 85, 86
electrolyte, vii, 1, 2, 17, 74, 76, 77, 85, 86, 88
electromagnetic, 43
electron, viii, ix, 3, 9, 11, 17, 18, 19, 20, 22, 29, 30, 31, 32, 33, 36, 37, 39, 43, 44, 48, 51, 61, 78, 79, 80, 84, 96, 106, 113, 114, 116, 131, 134, 135, 163, 164, 180, 203, 208
electron diffraction, 61
electron microscopy, ix, 3, 19, 20, 61, 113, 116, 131, 134
electronic structure, 49
electrons, 30, 31, 32, 33, 36, 43, 48, 78, 79, 146, 149
electrophoresis, 9, 20, 26, 181
electroplating, viii, 93, 96, 97, 99, 101, 104, 105, 106, 109, 111, 112, 168
embryonic stem cells, 168, 189
emission, ix, 37, 38, 39, 44, 48, 113, 146, 181, 182, 187, 203
encapsulation, 26
endothelial cells, 178, 193
energy, ix, 32, 33, 36, 38, 39, 43, 44, 50, 51, 52, 56, 57, 61, 62, 113, 119, 132, 147, 148, 149, 150, 151, 152, 156, 174, 181, 192
energy density, 56
engineering, 22, 30, 185, 200
environment, viii, 14, 55, 56, 57, 58, 63, 69, 157, 167
environmental change, 159
environmental impact, 30
environments, 20, 25, 114, 125, 176, 197
enzyme(s), 7, 8, 16, 17, 24, 157, 158, 159, 165, 172
epinephrine, 20
epitaxial growth, 116
equilibrium, 11, 36, 149
equipment, 56, 57, 58
ester, 18, 132, 160
etching, 16, 21, 22, 56, 70, 71, 211, 212, 213
ethanol, 4, 121, 161, 190
ethylene, 36, 78, 88, 165, 181
ethylene glycol, 78, 88, 165, 181
evaporation, 60, 172
evidence, 205
evolution, 9, 12, 13, 74, 78, 80, 83, 87, 90
excitation, 43, 48, 116, 134, 136, 137, 169, 171, 181, 182
excretion, 183

experimental condition, 80, 83, 84, 85
exploitation, 22
explosives, 167
exposure, 20, 159, 168, 211, 214
extinction, 42, 44
extracellular matrix, 163, 179

F

fabrication, vii, 16, 23, 24, 29, 163, 181, 192, 193, 199, 215
FEM, 58
fiber, 20, 21, 27
fibrinogen, 165
fibroblast proliferation, 176
fibroblasts, 165, 168, 178, 188
filament, 21, 75, 114, 115, 163
film thickness, 94, 114, 176
financial, 69, 90, 142
financial support, 69, 90, 142
flexibility, 56, 168, 176
fluctuations, 47, 48
fluorescence, x, 155, 157, 158, 168, 169, 171, 177, 181, 182, 183, 208, 211, 213
fluorine, 30, 38, 46, 49, 159, 165
force, 10, 56, 59, 94, 147, 149
formation, 6, 7, 12, 20, 31, 37, 46, 74, 82, 83, 86, 88, 118, 127, 133, 134, 142, 151, 184, 196, 198, 200, 204, 205, 210, 212, 215
formula, 42, 65
fouling, 2, 7, 9, 12, 14, 15, 20, 80, 86
fragments, 193
free energy, 61
Fresnel coefficients, 39
friction, vii, viii, 29, 30, 37, 38, 55, 56, 57, 58, 59, 62, 64, 65, 66, 67, 69, 71, 72, 111, 118, 120, 156, 173, 175, 176, 177, 192
FTIR, 47
fuel efficiency, 175
functionalization, 16, 18, 26, 186, 189, 193

G

gallium, 137, 138, 139
ganglion, 166, 179
gas diffusion, 191
gene expression, 165, 168, 170, 188
gene therapy, 171
genes, 172
genetic information, 9
geometry, 115
glasses, 30

glia, 171
glow discharge, 203
glucocorticoid, 165
glucose, 7, 8, 10, 15, 21, 24, 26, 27
glucose oxidase, 7
glutathione, 12, 26, 171
glycerol, 75, 88
glycol, 75, 79, 88, 89, 176, 182, 192
gold nanoparticles, 11, 14, 15, 25, 26, 171
gout, 11
grain boundaries, 5, 18, 68, 135, 137, 140, 142
grain size, 2, 63, 64, 134, 140, 141, 142, 162, 163, 208
graph, 125, 169
graphene sheet, 33
graphite, 30, 33, 34, 35, 36, 37, 67, 68, 118, 134, 146, 147, 167, 172, 173, 181, 189, 190, 196, 198, 204, 205, 210, 215
gratings, 116
growth, ix, x, 2, 5, 37, 69, 104, 106, 111, 113, 114, 116, 117, 120, 122, 124, 127, 132, 133, 152, 155, 162, 163, 164, 165, 169, 171, 173, 178, 179, 187, 188, 191, 193, 195, 196, 198, 199, 203, 204, 210, 211, 215
growth factor, 164
growth mechanism, x, 37, 162, 195
growth rate, 104, 106, 116, 117, 120, 122, 124, 127, 133, 163, 196, 211
guanine, 9, 13

H

hardness, vii, 29, 30, 34, 37, 38, 68, 114, 118, 163, 167, 169, 172, 173, 175, 197
heart attack, 170
heat loss, 59
heavy metals, 157
height, 8, 65, 115
helium, 174
heme, 9, 17, 18
heterogeneity, 39
hexane, 161
hip replacement, 176
histamine, 4, 5, 25
host, 148, 151
HRTEM, 61, 134, 138, 139, 142, 180
human, 9, 12, 157, 169, 179, 181, 185, 188, 193
humidity, 197
hybrid, 37
hybridization, vii, 29, 30, 31, 32, 33, 34, 37, 52
hydrocarbons, 34
hydrofluoric acid, 46
hydrogels, 186, 190

hydrogen, vii, x, 1, 2, 3, 6, 7, 8, 9, 10, 11, 14, 16, 19, 22, 24, 30, 31, 34, 35, 47, 49, 74, 80, 83, 114, 115, 127, 133, 135, 158, 159, 160, 162, 163, 164, 165, 172, 173, 174, 175, 186, 195, 198, 210, 212, 213
hydrogen atoms, 35, 159
hydrogen bonds, 158
hydrogen gas, 3
hydrogen peroxide, 14, 24, 165
hydrolysis, 9, 166
hydroperoxides, 12
hydroquinone, 27, 80, 81, 82, 83
hydroxide, 158
hydroxyl, 74, 80, 83, 90, 160, 161, 171
hydroxyl groups, 80, 160, 161, 171
hyperplasia, 176
hyperuricemia, 11

I

ICE, 76
ID, 134, 136, 141
identification, 134, 151
ileum, 23, 25
image, 77, 96, 97, 137, 138, 158, 181, 207
image analysis, 97
images, 3, 18, 19, 20, 77, 116, 134, 137, 138, 142, 169, 180, 182, 183, 198, 199, 201, 202, 204, 205, 206, 208, 209, 211, 213
immersion, 176, 193
immobilization, 2, 15, 16, 25, 27, 186
immune response, 166
immunofluorescence, 164, 179
impact energy, 148, 150
impact strength, 214
implants, 156, 165, 167, 177, 188
improvements, 174, 209
impurities, 121, 132, 133, 140, 147, 159
in vitro, x, 20, 156, 157, 169, 175, 176, 178, 179, 180
in vivo, x, 10, 20, 21, 26, 27, 125, 155, 156, 157, 164, 165, 168, 169, 170, 171, 175, 176, 177, 180, 182, 183, 184, 185
incidence, 39, 40, 116, 165, 169, 176, 177
incubation period, 210
indentation, 200, 201, 205, 206, 209, 212, 213
inhomogeneity, 124, 159
initial state, 43
insulin, x, 155, 157, 170, 184
integrated circuits, 120, 125
integration, 149, 165
integrin(s), 163, 164
integrity, viii, 55, 57, 69, 166

interface, vii, viii, 1, 2, 18, 22, 39, 40, 55, 57, 58, 59, 60, 61, 64, 71, 166, 177, 190, 198, 199, 200, 204, 205, 210, 212
interfacial adhesion, 207
interfacial bonding, 212
interfacial layer, 196
interference, 12, 135, 168
intestine, 21
intravenously, 182
inversion, 146
iodine, 30, 38
ion implantation, ix, 145, 147, 151, 176, 193
ions, 32, 33, 86, 119, 121, 137, 138, 152, 176
IR spectra, 49
iridium, 14, 26
iron, 56, 61, 70, 94, 120, 198
irradiation, 112, 151, 194
isomers, 31, 80, 81, 83
isotope, 147
issues, 171, 207

K

kidney(s), 6, 169, 183
kinase activity, 23
kinetics, 18
KOH, 15

L

labeling, 189
lactic acid, 169
laminar, 146, 147
laser ablation, 112, 180, 181
lead, 18, 80, 167, 168, 177
lesions, 5, 176
lifetime, 48, 168, 197
light, 39, 101, 125, 182
linear defects, 149
linear function, 8
lipids, 170
liquid chromatography, 12, 13, 23
lithography, 23, 125
liver, 5, 169, 183
liver cancer, 169
liver damage, 5
longevity, 156, 166, 176, 177, 197
love, 113
low temperatures, 163
lubricants, 168
luciferase, 157, 158
luminescence, 181

Luo, 21, 24, 129, 193, 194
lying, 33, 48
lymph, 182
lymph node, 182
lymphocytes, 25
lysine, 18, 158, 178, 179
lysozyme, 157, 158

M

machinery, viii, 55, 69
macular degeneration, 166
magnitude, 4, 13, 22, 38, 132, 140, 141, 142, 175
manganese, 70, 198
mapping, 67, 213
mass, 31, 77, 83, 121, 124, 132, 167, 186
mass spectrometry, 132
materials, vii, ix, x, 1, 2, 3, 6, 11, 16, 22, 29, 30, 31, 35, 37, 38, 43, 44, 56, 64, 68, 69, 74, 90, 114, 120, 140, 141, 147, 151, 155, 156, 157, 158, 162, 165, 167, 170, 171, 174, 177, 184, 186, 191, 195, 196, 197, 198, 203, 207, 210, 214, 215
matrix, vii, 26, 29, 30, 36, 38, 39, 48, 51, 52, 122, 160, 169, 198
matter, iv, 35, 74, 81, 82
measurement(s), ix, 8, 9, 10, 12, 20, 21, 25, 26, 27, 38, 45, 58, 59, 60, 65, 72, 75, 76, 86, 122, 127, 131, 135, 140, 141, 142, 170, 177, 213, 214
mechanical properties, vii, 29, 30, 38, 39, 173, 176, 191, 197, 198, 210, 214
media, 14, 30, 38, 39, 77, 78, 88, 164, 165, 171, 181
MEMS, 126, 174
mesenchymal stem cells, 165, 179
metabolism, 157
metabolites, 2
metal ion(s), viii, 93, 96, 177
metal nanoparticles, vii, 1, 6, 15
metal oxides, 62, 74
metals, x, 30, 37, 56, 62, 70, 157, 167, 195, 196, 198, 203, 210, 215
methanol, 46, 76, 174
methodology, 14, 147
methylene blue, 78
microcrystalline, 77, 106, 163
microelectronics, 165
microscope, 51, 76, 106, 180
microscopy, ix, x, 75, 131, 133, 134, 155, 157, 169, 171, 181
microstructure, 116, 133, 142, 187, 210, 214
microstructures, 197, 210, 212
mineralization, 75, 89
miniaturization, 7
modelling, viii, 55, 57, 58, 60

models, 169, 182, 210
modifications, 135
modulus, 65, 169, 174, 176, 196, 198
molecular beam, vii, 145
molecular dynamics, ix, 145, 146
molecular oxygen, 16
molecular weight, 177, 192
molecules, vii, 2, 7, 9, 10, 12, 14, 16, 17, 29, 31, 79, 157, 158, 159, 161, 163, 165, 167, 172, 177, 180, 193
molybdenum, 134
momentum, 43, 44
monolayer, 178, 184
morphology, 77, 116, 133, 134, 142, 162, 171, 180, 203
mucosa, 20, 27
multiwalled carbon nanotubes, 189, 194
myoglobin, 17

N

Na2SO4, 75, 79, 86, 87
NaCl, 75, 86, 87, 88, 89
NADH, 3, 4, 10, 23, 25
nanocrystals, 186, 194
nanodots, x, 155, 157, 180, 181
nanofibers, 2, 26, 186
nanoimprint, 23
nanometer, 18, 37, 188
nanometer scale, 37
nanometers, 94
nanoparticles, x, 14, 15, 27, 155, 157, 159, 160, 161, 168, 169, 170, 171, 180, 181, 183, 184, 186, 189, 194
nanorods, 156, 185
nanostructures, vii, 1, 3, 19, 22, 203
nanotube, 22, 27, 193
nanotube films, 22
nanowires, 19, 22, 25
National Academy of Sciences, 189, 190
nervous system, 20, 164
Netherlands, 76
neuronal cells, 164, 179, 191
neurons, 10, 171, 179, 193
neurotransmission, 20
neurotransmitter, 10, 167
nickel, viii, 7, 15, 26, 57, 93, 94, 96, 97, 104, 105, 106, 107, 108, 109, 110, 111, 112, 176, 198
nicotinamide, 3
NIR, 38
nitric oxide, 179
nitrides, 203

Index

nitrogen, ix, 30, 38, 63, 78, 132, 145, 147, 150, 151, 157, 168, 174, 191
nitrogen gas, 63, 78
nodes, 74
non-metals, 196
norepinephrine, 20, 24, 25, 167
NRC, 215
nuclear membrane, 182
nucleation, x, 116, 122, 124, 127, 162, 163, 188, 195, 196, 198, 199, 202, 203, 206, 207, 209, 210, 211, 212, 215
nuclei, 33, 170, 171, 181, 182
nucleic acid, 9, 24, 25
nucleus, 31, 171

O

oil, 45, 175
oligodendrocytes, 164
oligonucleotide arrays, 177, 178
oncogenesis, 190
opportunities, 23, 177
optic nerve, 166
optical microscopy, 116
optical properties, vii, viii, 29, 30, 38, 39, 44, 51
optimization, 202, 213
optoelectronic properties, 36
organic compounds, 74, 75, 78, 88, 124
organic matter, 74, 80
organic solvents, 38
oxidation, viii, 3, 4, 5, 6, 7, 8, 9, 10, 11, 12, 13, 14, 15, 16, 18, 19, 20, 21, 24, 25, 26, 30, 56, 62, 63, 70, 73, 74, 75, 76, 79, 80, 81, 83, 84, 85, 86, 87, 88, 89, 112, 160, 161, 177, 180, 181, 187, 194, 202
oxidation products, 3, 19
oxidative stress, 12, 171
oxygen, vii, 1, 2, 3, 6, 7, 9, 10, 11, 12, 13, 14, 15, 19, 21, 22, 62, 63, 68, 74, 75, 76, 78, 80, 83, 87, 90, 164, 165, 167, 175, 181
oxygen plasma, 21, 22

P

p53, 168
parallel, 45, 149
passivation, 181, 182
pathways, 18, 39, 74, 157, 164
PCR, 164, 170
peptides, 159, 182
pericytes, 193
permeability, 175

permission, iv, 90, 158, 160, 162, 164, 169, 170, 178, 179, 180, 182, 183
peroxide, 171
PET, 175, 191
pH, 4, 5, 6, 8, 12, 13, 15, 17, 18, 21, 77, 79, 80, 81, 82, 86, 87, 88, 170, 184
phase diagram, 34, 172
phase transformation, 58, 61, 63, 210
phenol, 17, 24, 25, 80, 83, 87
phenolic compounds, 17, 87
phenotype, 164
phonons, ix, 44, 145, 151
phosphate, 4, 81, 82, 87, 88
phosphates, 86
phosphorous, 30, 38, 156, 177, 178
phosphorus, 178, 190, 192, 193
photobleaching, x, 156, 157, 168, 169, 182, 183, 184, 190
photoelectron spectroscopy, 60
photoluminescence, 30, 48, 49, 181
physical properties, 165, 170, 215
PL spectrum, 49
plasma proteins, 177, 178
plasmid, 171
platform, 17, 165, 180
platinum, 7, 74, 76, 81, 82, 83, 84, 87, 88, 89, 171, 185
PM, ix, 145, 148, 149, 151
point defects, 149, 157, 168
polar, 2, 20, 158
polarity, 165
polarization, 5, 39, 48, 74, 80, 86
pollutants, viii, 73, 75, 89
polyamine, 5
polyamines, 5, 24
polyimide, 167
polymer, 11, 75, 82, 83, 121, 169, 196
polymer matrix, 169
polymerase, 164
polymerase chain reaction, 164
polymerization, 16, 19
polymers, 34, 74, 83, 88, 177
polymethylmethacrylate, 121, 124
polypeptide, 9
polyphenols, 25
polystyrene, 15, 26, 164
porosity, 30, 212
potassium, 94
pressure gauge, 45
probability, 31, 43, 132, 151
probe, ix, 131, 133, 134, 167, 168
product performance, 197
profilometer, 100, 107

progenitor cells, 182, 190
pro-inflammatory, 165
project, 189, 199
proliferation, 164, 165, 169, 171, 178, 179, 188
proposition, 181
prostheses, x, 155, 156, 167, 184
prosthesis, 166
protection, 18, 38
protective coating, viii, 29, 37, 114, 196
protein structure, 159
proteins, 8, 9, 17, 18, 158, 159, 165, 166, 168, 172, 179, 182, 186
pumps, 45
purification, 160, 172, 178, 180, 186, 193
purines, 9, 26
purity, 47, 121, 159, 203, 206, 207, 208
pyrimidine, 13, 23

Q

quality of life, 156
quantification, 13
quantum computer, ix, 145, 147, 151
quantum confinement, 181
quantum dot(s), x, 156, 157, 168, 180, 181, 193
quantum entanglement, 147
quartz, 45, 76, 196
quasi-equilibrium, 148, 149
quasi-equilibrium state, 148, 149
qubits, 147
quinone(s), 16, 83

R

radiation, vi, x, 116, 125, 137, 148, 150, 159, 168, 181, 195, 215
radiation damage, 148, 150
radical mechanism, 16
radicals, 5, 74, 83, 86, 90, 159, 161, 171, 187
radiotherapy, 125
radius, 15, 22, 31
Raman spectra, 20, 61, 116, 119, 120, 123, 124, 135, 136, 137, 200, 201, 205, 206
Raman spectroscopy, ix, 61, 113, 116, 120, 124, 127, 131, 133, 134, 137, 141, 142, 191
reactants, 210
reactions, vii, 1, 2, 15, 16, 17, 56, 58, 61, 151, 161, 177, 181
reactive groups, 16
reactive oxygen, 171
reactive sites, 21, 159
reactivity, x, 81, 84, 195, 198

reagents, 75
recovery, 30, 151, 152, 170
refractive index, x, 40, 41, 44, 155, 175
refractive indices, 39, 41, 42
researchers, vii, 1, 137, 176, 196
residues, 9
resilience, x, 156
resistance, x, 2, 30, 106, 119, 120, 155, 161, 175, 176, 177, 183, 191, 197, 200, 210
resolution, ix, 26, 51, 116, 131, 134, 138, 180, 213
resorcinol, 24, 80, 81, 82, 83
resources, 183
response, vii, viii, 1, 2, 4, 6, 7, 8, 12, 13, 15, 17, 18, 20, 21, 22, 23, 24, 73, 151, 176, 192
restenosis, 176
restrictions, 174
retina, 166
retinitis, 166
retinitis pigmentosa, 166
rings, 36
room temperature, 47, 48, 58, 116, 121, 132, 147, 151, 174, 180, 196, 207
roughness, x, 56, 59, 63, 64, 65, 67, 68, 93, 94, 96, 100, 101, 108, 110, 111, 116, 155, 163, 166, 175
routes, x, 155, 157, 159, 160, 172, 174, 180
Royal Society, 143
rules, 43

S

salmonella, 166
salts, 159, 181, 187
sapphire, 196
saturation, 120
scaling, 149
scanning electron microscopy, ix, 3, 18, 131, 133, 134
scattering, 37, 150
school, 145
science, 70, 152
scientific method, 147
scope, 75, 120
seeding, ix, 113, 117, 211
segregation, ix, 131, 138, 139, 142
selectivity, 5, 7, 10, 12, 14, 15, 19, 21, 22, 156, 165
self-assembly, 14, 15
SEM micrographs, 97, 101, 102, 105, 106, 109, 110, 117, 124
semiconductor, ix, 43, 44, 50, 131, 132, 140, 145, 146, 193
semiconductors, 36, 48, 50
sensing, 14, 73

sensitivity, vii, 1, 2, 3, 4, 5, 6, 10, 12, 13, 14, 15, 17, 19, 20, 21, 22, 125, 168
sensors, 7, 10, 14, 16, 18, 26, 166
serotonin, 4, 20, 23, 25, 27
serum, 12, 159, 164
serum albumin, 159
shape, 77, 127, 135, 174, 176, 210
shape-memory, 176
shear, 175, 209
shear deformation, 209
shear strength, 175
shock, 167
showing, 7, 115, 124, 137, 169, 171, 200
Si3N4, 196
side effects, 170
signalling, 163
signals, 9, 167
signs, 106, 171
silane, 161
silicon, ix, 36, 46, 47, 48, 49, 95, 113, 114, 116, 117, 119, 120, 127, 131, 132, 133, 134, 147, 156, 163, 165, 166, 171, 177, 187, 190, 192, 198
silver, 30, 47, 51
SiO2, 196
smoothness, 18, 207, 209
sodium, 77, 78, 81, 82, 88, 170
sodium hydroxide, 77, 170
solubility, x, 118, 195, 196, 210
solution, viii, 2, 3, 4, 8, 11, 12, 13, 15, 16, 18, 19, 21, 46, 73, 74, 76, 77, 83, 84, 85, 86, 88, 121, 124, 159, 170, 182, 196, 198
Soviet Union, 162
Spain, 194
spatial location, 11
species, 2, 3, 5, 10, 16, 23, 74, 147, 171, 174, 180, 210, 212
spectrophotometry, 75
spectroscopy, ix, 61, 113, 116, 125, 131, 170, 211, 213
stability, vii, 1, 2, 3, 5, 7, 10, 11, 12, 13, 14, 16, 17, 18, 20, 21, 22, 30, 74, 86, 133, 156, 159, 165, 166, 183
standard deviation, 14
steel, 39, 45, 57, 77, 114, 116, 118, 119, 120, 121, 127, 167, 176, 191, 192, 197, 198, 199, 200, 202, 203, 204, 207, 208, 209, 210, 215
stem cells, 164, 179, 187, 188, 189
stent, 176, 192
storage, viii, 16, 29, 37, 175
storage media, 175
stress, x, 30, 38, 52, 121, 124, 127, 134, 136, 163, 167, 195, 204, 208, 209, 211, 213
structural changes, 215

structural defects, 133
sulfur, 5
sulfuric acid, 84, 85
superconductor, 132
suppression, 10, 19
surface area, 64, 103, 163, 167, 174, 215
surface chemistry, 30, 168, 187
surface energy, 181
surface hardness, 197, 200
surface layer, 119, 161, 165, 167, 204
surface modification, 22, 156, 157, 158, 159, 161, 215
surface properties, 163, 164, 188, 197
surface region, 205, 214
surface structure, vii, 1, 2, 3
surface tension, 177, 178
surface treatment, 114, 116, 121, 122, 167, 192
surfactants, 75, 79, 80, 86
symmetry, 146, 149
sympathetic nervous system, 25
synthesis, iv, vii, viii, x, 29, 46, 47, 55, 75, 156, 163, 167, 178, 189, 193, 194, 198, 215

T

tantalum, 198
target, 47, 150, 151
technical assistance, 127
techniques, viii, ix, 2, 12, 20, 55, 56, 58, 68, 69, 90, 112, 114, 118, 131, 159, 162, 163, 174, 176, 196, 200
technology(ies), viii, 17, 22, 38, 55, 57, 69, 72, 114, 147, 152, 175, 180, 203, 215
teeth, 176
TEM, ix, x, 61, 131, 133, 134, 139, 142, 170, 180, 195, 202
temperature, viii, ix, 30, 37, 38, 39, 45, 46, 47, 48, 55, 56, 57, 58, 59, 60, 61, 63, 64, 65, 66, 69, 70, 71, 72, 77, 94, 114, 121, 124, 131, 133, 138, 140, 141, 145, 148, 149, 151, 156, 162, 163, 174, 187, 190, 196, 203, 211, 215
tension, 45
TEOS, 46, 47
therapy, 171, 188
thermal activation, 37
thermal decomposition, 172
thermal deformation, 66
thermal energy, 56, 175
thermal expansion, x, 66, 68, 195, 196, 198, 207, 210, 213
thermal properties, 64
thermal stability, 30
thermal treatment, 120

Index

thermodynamics, 61
thermogravimetric analysis, 121
thin films, vii, viii, 1, 2, 3, 14, 22, 23, 26, 29, 30, 44, 46, 47, 68, 69, 159, 166, 178, 184, 186, 188, 189, 193, 196, 200, 207
thymine, 13
tissue, 20, 25, 125, 165, 167, 169, 176, 177, 184
tissue engineering, 169, 184
titanium, 30, 57, 60, 63, 176, 188, 193, 198, 210, 212, 214
toxicity, ix, 131, 168, 169, 170, 171
transformation(s), 61, 62, 63, 72, 84, 167
transistor, 30
transition metal, x, 195, 197, 198, 204, 207, 210, 215
translation, 180
transmission, ix, 40, 41, 42, 43, 47, 51, 61, 131, 133, 134, 138, 170, 180
transmission electron microscopy, ix, 61, 131, 133, 134, 138, 170
transmittance spectra, 49, 51
transparency, viii, 29, 30, 37, 38, 47, 114, 163, 175
transport, 9, 38, 77, 83
treatment, 3, 7, 10, 11, 21, 74, 88, 89, 120, 122, 134, 159, 169, 170, 176, 187, 197, 202, 203, 211, 215
tryptophan, 8, 22, 23, 25, 27
tumour growth, 169
tumours, 169
tungsten, ix, 39, 113, 114, 198
tunneling, 38
twinning, ix, 131, 133, 138
twist, 197
tyrosine, 8, 23, 27

U

uniform, 58, 66, 115, 121, 163, 198
uric acid, 7, 9, 24, 25
urine, 6, 12, 182, 183
UV, 37, 38, 44, 75, 76, 174, 177, 181, 182, 190
UV irradiation, 177
UV spectrum, 181

V

vacuum, 14, 37, 45, 56, 57, 94, 114, 120, 125, 174, 203

valence, 32, 43, 44, 132
vapor, 14, 69, 70, 75, 112, 132, 187, 210
variables, 64, 65
variations, 59, 165
vector, 39
vehicles, x, 155, 158, 161, 169, 170, 172, 184
velocity, 65, 149
vibration, 149, 150, 151
video microscopy, 20, 24
viruses, 170
vision, 166
vitamin E, 12

W

wastewater, 74
water, 7, 74, 75, 76, 79, 80, 83, 86, 89, 121, 166, 167, 170, 180
water supplies, 166
wave vector, 44
wavelengths, 41, 42, 157, 182
wear, viii, x, 29, 30, 37, 38, 106, 111, 118, 120, 156, 175, 177, 184, 188, 192, 195, 196, 197, 200, 210
Western blot, 171
wettability, 192
wide band gap, 163
windows, viii, 29, 37, 114, 125
wires, 176, 185, 192
workers, 39

X

XPS, 47, 51, 166
X-ray analysis, 213
X-ray diffraction (XRD), ix, 61, 113, 116, 117, 131, 133, 134, 161, 181

Z

zinc, 182
zirconia, ix, 131, 133, 134, 137, 138, 141, 142
zirconium, 134, 198